THE SILVER BURDETT ENCYCLOPEDIA OF TRANSPORT

AIR

J.D. Storer

Published 1983 by The Hamlyn Publishing Group Limited
Astronaut House, Feltham, Middlesex, England
© Copyright The Hamlyn Publishing Group Limited 1983
All rights reserved. No part of this publication may be
reproduced, stored in a retrieval system, or transmitted,
in any form or by any means, electronic, mechanical,
photocopying, recording or otherwise, without the
permission of the Hamlyn Publishing Group Limited.

Adapted and published
in the United States by
Silver Burdett Company
Morristown, N.J.
1984 Printing
ISBN 0-382-06775-4 Lib. Bdg.
ISBN 0-382-06990-0
Library of Congress
Catalog Card Number 83-51807

Photographs
Air France, London 20 bottom; Associated Press, London 21 bottom; Australian Information Service, London 49 bottom, 56 bottom; BBC Hulton Picture Library, London 11 bottom; British Aerospace, Hatfield 15 top, 15 bottom, 29; British Airways, London 9 bottom, 16, 16 inset, 17 bottom, 18–19, 19, 20 top, 32–33, 36; Central Press, London 21 top; Daily Telegraph Colour Library, London 56 top; Goodyear Tire & Rubber Co., (Great Britain) Ltd. 7; David Halford 18 bottom, 30 top, 37 top, 37 bottom, 64; Hamlyn Group Picture Library 4–5, 17 top, 24, 45 top, 45 bottom, 49 top, 62 top; High Commissioner for New Zealand, London 48; Illustrated London News Picture Library, London 6, 11 center; Popperfoto, London 41; RNLI — R.J. Wilson 46; Renegade, Wimborne 58 left; Rex Features, London 3 top, 42–43, 43; Rolls-Royce, Derby 12, 27 top; Royal Scottish Museum, Edinburgh 12; Science Museum, London 8, 9 top, 10, 12; Scott Polar Research Institute, Cambridge 3 bottom; Brian M. Service, Ashford 26–27, 34–35, 50–51; Smiths Industries, Cheltenham 58 right; Spectrum Colour Library, London 1; Michael Taylor, Cheam 44 top, 44 bottom, 57; US Army AAF 25; ZEFA (UK), London — Robert Lorenz 30 bottom, Richard Nicholas 59.

Illustrations
Derek Bunce; David Lewis Management; Linden Artists Ltd.; Gerald Whitcomb; John W. Wood.

AIR

Balloons	2	Airways	32
Airships	4	Airports	34
Birth of the airliner	8	World airlines	38
Trail-blazers	10	Freight aircraft	40
Power to fly	12	Flying boats	42
How aircraft fly	14	Helicopters	46
Air travel 1919–1959	16	Help from the air	48
The jet age	20	STOL and VTOL aircraft	50
A modern airliner	22	Air rescue	55
Faster than sound	24	People at work	58
Building an airliner	28	Light aircraft	60
Flying an airliner	30	The future	62

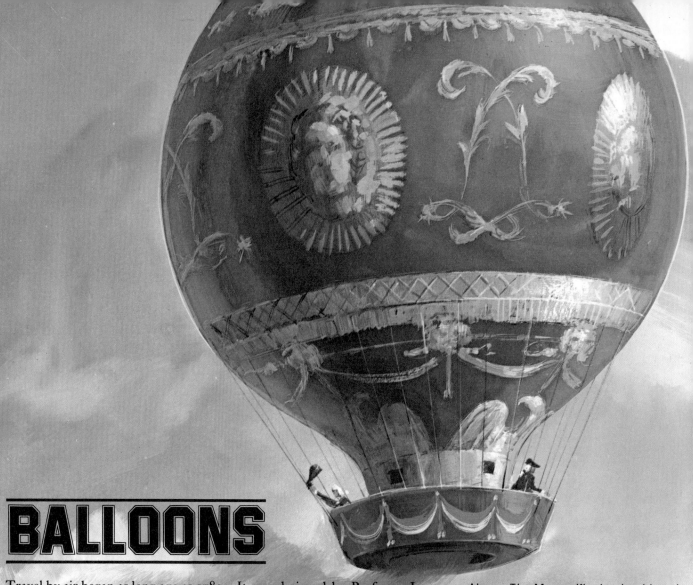

Above: *The Montgolfier brothers' hot air balloon.*

BALLOONS

Travel by air began as long ago as 1783 when two Frenchmen made a voyage across Paris in a large, highly decorated balloon filled with hot air. The idea of filling a balloon with hot air – which is much lighter than cold air – was developed by two brothers, Joseph and Etienne Montgolfier. They did not wish to fly, so the honor of the first aerial voyage went to Pilâtre de Rozier and the Marquis d'Arlandes. The Montgolfiers' balloon was open at the bottom and in this opening there was a fire carried in a brazier. On 21 November 1783 everything was ready and slowly the balloon filled with hot air: then majestically it lifted off the ground carrying the first balloonists on a historic $5\frac{1}{2}$ mile flight lasting about 25 minutes.

Two weeks later another balloon flight took place in France but this time the balloon was not filled with hot air. It was designed by Professor Jacques Charles, who was a scientist at the Paris Academy and therefore knew about the recently discovered gas – hydrogen – which is very much lighter than air. Professor Charles was assisted by the two Robert brothers and on 1 December 1783 the hydrogen balloon carried Professor Charles and the elder Robert on a 26 mile cross-country flight.

Ballooning became a popular sport but as a means of transport these craft had one great disadvantage: they were at the mercy of the wind. A balloon travels in whichever direction the wind happens to be blowing, and winds are notoriously difficult to predict. Nevertheless, some remarkable balloon flights were made. In January 1785, a French balloonist called Jean-Pierre Blanchard and an American scientist Dr John Jeffries took off from Dover, England in a hydrogen balloon. The wind was blowing towards France but after a good start their balloon sank lower and lower. They threw everything they could spare overboard, even their clothes, and the balloon rose again. Blanchard and Jeffries reached France safely – but very cold! Several scientists followed Dr Jeffries' lead and used balloons to explore the unknown atmosphere. The British pair, Coxwell and Glaisher made many research flights and even tried to take aerial photographs. On one of their ascents in 1862 they reached a height of over 25 000 feet – though this was more by accident than design, because a control valve stuck.

Balloons had been pressed into military service as tethered look-out posts in 1794, but in 1870 a new military use was introduced during a war between France and Germany. Paris was completely surrounded by German troops and because there were no telephones or radio in 1870, the city was completely isolated. Several balloonists trapped in Paris suggested making flights over the enemy lines to the regions still held by French troops. The daring plan worked: passengers and letters were carried by balloons from the city – often at night. Of course the wind had to be blowing in the right direction and if this direction changed during a flight the balloonist was in trouble. Homing pigeons were carried in the balloons to take communications back to Paris; messages on microfilm (very small photographs) were attached to the pigeons and they flew back to their lofts in Paris. This remarkable operation lasted from September 1870 to January 1871.

Ballooning in the 20th century

During the early years of the present century, ballooning became a popular sport for wealthy amateurs, who indulged in games and competitions. The balloons were usually filled with hydrogen or coal gas (which was much cheaper) and the aeronauts (as balloon travelers were called) were carried in a wicker basket. By 1910 the excitement of ballooning was superseded by the thrill of flying in powered aircraft, with the result that balloons almost vanished from the skies. In the 1930s a small number of research balloons were produced to explore the upper atmosphere. One of the most famous scientists to use a balloon for this work was Professor Auguste Piccard of Switzerland. He made the first flight into the cloud-free, extremely cold, upper layer of the Earth's atmosphere called the *stratosphere* in 1931. He and Paul Kipfer ascended to a height of over 52 000 feet.

A revival of ballooning as a sport began in the 1960s with the introduction of cheap and reliable hot-air balloons. These work on the same principle as the Montgolfier hot-air balloon but, instead of using a brazier to heat the air, they have a powerful gas burner which is supplied with propane gas from pressurized cylinders. The balloon's *envelope* is usually made from special nylon but the *skirt* through which the flame passes has to be fire resistant, so asbestos and glass fibre are often used.

One of the most remarkable balloon flights in recent years took place in 1978 when three balloonists from the United States made the first balloon crossing of the Atlantic Ocean. Their balloon, called *Double Eagle II*, was filled with the lightweight gas *helium* and it ascended from Maine, USA, on 11 August. The westerly wind carried it across the North Atlantic, over Ireland and southern England to land in France – about 60 miles short of Paris. The journey lasted six days and the distance traveled was over 1800 miles – a far cry from Pilâtre de Rozier's and the Marquis d'Arlandes' $5\frac{1}{2}$ mile flight across Paris in 1783.

Above: *The first Atlantic balloon crossing, made by* Double Eagle II *in 1978.*

An ill-fated flight

One of the most ambitious balloon flights of the 19th century was the attempt by the Swedish engineer Salomon Andrée to reach the North Pole. In 1897 he set out with two companions from one of the Spitsbergen islands to the north of Norway, and they drifted over the ice towards the Pole. Unfortunately, ice formed on the hydrogen-filled balloon and the crew had to dump overboard all the ballast and everything else which could be spared. Despite their efforts the balloon came down on the ice after being airborne for 65 hours, yet they were still not halfway to the Pole. Andrée and his crew set out on the long trek back to civilization but after two and a half months they perished. Their fate was not known until 1930 when their last camp was discovered. Amongst the remains were diaries, and some undeveloped films taken by the explorers: these were carefully processed and remarkable photographs of the ill-fated expedition emerged.

Below: *This photograph of Andrée's balloon was found 33 years later.*

AIRSHIPS

Soon after balloons were invented, enterprising balloonists attempted to control the direction of their flights by operating large paddles or oars. Unfortunately, these had no effect and the balloons still went wherever the wind took them. To make a dirigible (steerable) balloon or airship some form of lightweight engine, driving a propeller, was needed. But, until the late 19th century, the only type of engine available to airship designers was the steam engine and this was particularly heavy because it needed a boiler to raise steam. Nevertheless a steam-powered airship was built in 1852 by the French engineer, Henri Giffard. It consisted of a sausage-shaped balloon some 130 feet long and suspended beneath this was a steam engine driving a large propeller. A triangular rudder was fitted at the rear to steer the craft like a ship. At full power this primitive airship could reach a speed of about 5 mph (8 kph) – a brisk walking pace – so it could not cope with a strong headwind, but it was a step in the right direction.

During the latter half of the 19th century two new sources of power emerged: the electric motor, running off batteries, and the gasoline engine. In 1884 Charles Renard and Arthur Krebs built an airship in France, which they called *La France*, and this was powered by an electric motor. Traveling at over 12 mph (20 kph), *La France* could be controlled reasonably well in flight, and it even completed several round trip flights. However, the weight

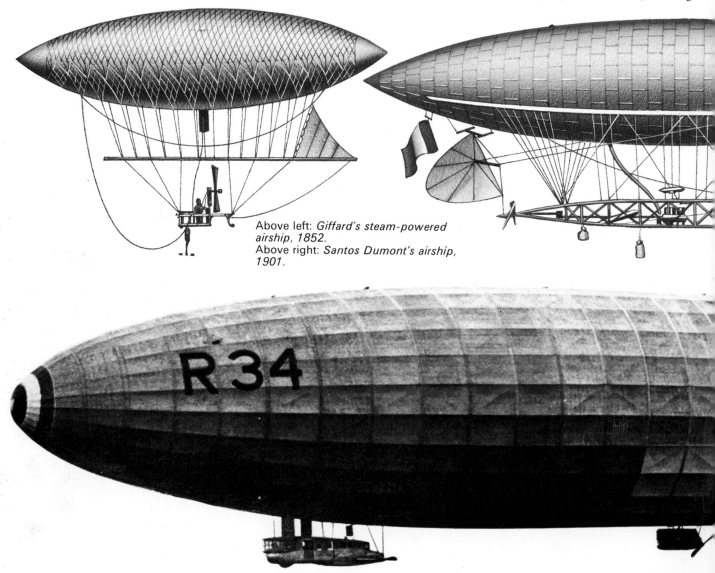

Above left: *Giffard's steam-powered airship, 1852.*
Above right: *Santos Dumont's airship, 1901.*

of its batteries proved to be rather a handicap. By 1886 two German engineers, Gottlieb Daimler and Karl Benz, had both invented gasoline engines, and this gave hope to airship designers. The first airship with a gasoline engine was built in 1888 by another German, Karl Wölfert, but this was not a great success. The man who really captured the imagination of the public with his airships was Alberto Santos Dumont – a Brazilian who lived in Paris. His most notable flight took place in 1901 when he flew from St Cloud to the Eiffel Tower and back in less than half an hour, thereby winning a large cash prize. The total distance was only 12 miles, however several other attempts failed because of headwinds.

Most of these early airships consisted of a large sausage-shaped balloon with a *car* suspended below to carry the crew and engine. They were known as *non-rigid* airships because of the flexible nature of the large gas bag. During the First World War, non-rigid airships played an important part by patrolling the seas in search of enemy ships.

Giant airships and blimps

Large, non-rigid airships were very unwieldy and the German airship designer, Count Ferdinand von Zeppelin, designed huge and powerful *rigid* airships. They had a lightweight metal framework and the gas was contained in a large number of separate gas bags fixed inside the fabric-covered framework. After a number of failures, Zeppelin was joined by Dr Hugo Eckener and by 1910 they had regular passenger services operated by airships or *Zeppelins*. During the four years prior to the First World War (1914–1918) these Zeppelins carried a total of 34 000 passengers without loss of life – although there were some narrow escapes.

Throughout the First World War, Zeppelins were in action, patrolling the seas and bombing targets in Britain. One of them was brought down almost intact in 1916 and British designers set to work designing an airship based on the crashed Zeppelin's design: the previous British airships had not been very successful. Two successful airships – the *R33* and the *R34* – were built to these plans but they were not completed in time to take part in the war. The *R34* made a historic flight in July 1919 when it completed the first non-stop east–west crossing of the Atlantic by air in 4.5 days (the first non-stop flight in the opposite direction took place a few weeks earlier by Alcock and Brown in a Vickers Vimy bomber). Not only did the airship *R34* fly direct from its base in Scotland to New York, but it also made the return journey to England a few days later, in only 3 days, with the help of a tailwind.

The *R34* was not able to carry passengers across the Atlantic but within 10 years a passenger-carrying airship was operating a trans-Atlantic service, and once again it was one of German design. The *Graf Zeppelin* first flew in 1928 and made many successful flights. During the following years several giant airships were built – all more than 650 feet long. The British built the *R100* and *R101*, the Americans produced the *Akron* and the *Macon*, and in

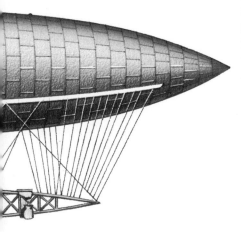

Above: *The British airship R34 at Pulham in England after completing a return flight across the Atlantic in 1919.*

Above: *As the German airship Hindenburg arrived at New York in 1937 it burst into flames. This was the end of the era of giant airships.*

1936 the Germans built the largest of them all, the *Hindenburg*. Of these six airships, four were destroyed in dramatic accidents with the loss of many lives: passenger-carrying airships were then abandoned.

Smaller non-rigid airships or *blimps* continued to be built in small numbers and during the Second World War (1939–1945) the United States Navy used blimps from 1941 onwards to escort convoys of ships. These slow-flying airships were ideal for spotting submarines or other enemy warships and it is claimed that no ship was sunk while it had a blimp escort. The Goodyear Tire and Rubber Company built many of the Navy blimps and continued to build airships when peace was restored – as airborne advertisements.

Will large airships return in the future? Predictions that airships will return as the cargo-carriers of the future have been made frequently over the past 20 years or more, yet so far little progress has been made. To attract cargo away from ships or aircraft the airships will have to be not just equal to existing services, they must be better all round.

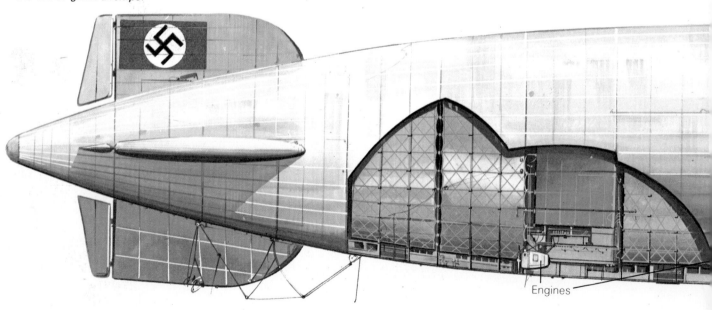

Above: *A modern non-rigid airship, the Goodyear* Europa *is used for joy-rides, advertising and carrying TV cameras.*

Below: *The gigantic German Zeppelin LZ129* Hindenburg *of 1936 carried 50 passengers in great luxury. They were housed inside the hull where there was even a dance floor. Four large diesel engines propelled the airship at almost 85 mph.*

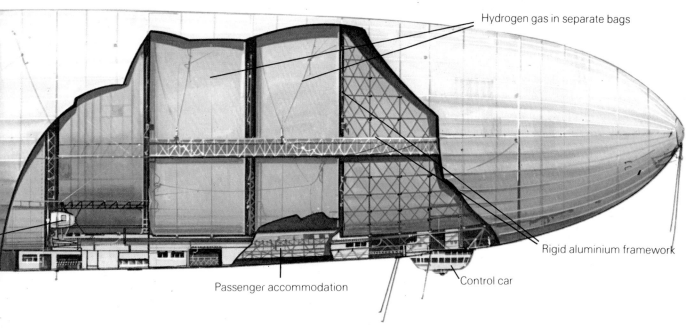

Hydrogen gas in separate bags

Rigid aluminium framework

Passenger accommodation

Control car

BIRTH OF THE AIRLINER

When man first tried to fly with wings it seemed natural to copy the birds and try to recreate their wing-flapping action, yet this was a false trail and no successful wing-flapping flying machine or *ornithopter* has ever been built. That great inventor of the 15th century, Leonardo da Vinci, designed several ornithopters – none of which could ever have worked. The answer was to use fixed wings – like a soaring bird. A British engineer called Sir George Cayley sketched a fixed-wing aircraft in 1799 and half a century later, he built a full-size glider which carried his coachman across a small valley. The coachman was a reluctant pilot and threatened to resign – he did not appreciate the fact that he had made the first gliding flight in history! Although Cayley studied the possibilities of powered flight, he lacked a suitable engine.

In 1842–3 two other British designers made a serious study of powered flight: William Samuel Henson and his friend John Stringfellow designed a magnificent *Aerial Steam Carriage*. Prints of the projected machine were produced, showing it in flight over London and even cruising over the Pyramids in Egypt. Lack of money prevented its construction but Stringfellow did make a steam-powered model which flew across a room.

Invention of the gasoline engine in 1886 made powered flight possible, but many of the early attempts were nothing more than uncontrolled hops: real flight required control through the air and, for this, man had to learn the art of gliding. After studying the soaring flight of birds a German engineer called Otto Lilienthal built his first glider in 1891. He controlled it by moving his weight about – modern hang-gliders use the same principle. After making some 2000 flights Lilienthal was killed in a gliding accident but his work was continued by the British pioneer Percy Pilcher who also died as the result of a crash. An American, Octave Chanute, collected together all the available information on gliding and built further gliders – although his most important contribution may have been the encouragement he gave to his fellow countrymen Wilbur and Orville Wright.

Powered flight

The Wright brothers set about solving the problems of flight in their own methodical way. First they developed a means of controlling a glider using movable control surfaces – instead of weight shifting. Then they built their own lightweight gasoline engine and on 17 December 1903 they made the world's first genuine flight in their biplane *Flyer*. Because they were worried in case rivals might steal their ideas, the Wrights did most of their flying in secret: in fact the attempts of another American pioneer Samuel Pierpont Langley received much more publicity – despite ending in failure. The original Wright *Flyer* made four flights, the longest lasting just 59 seconds. A new *Flyer* was built in 1904 and this made about 80 flights including

Below: *Otto Lilienthal flying his glider in 1896.*

simple maneuvers and a complete circuit over the field. The longest flight achieved was 5 minutes 4 seconds. During the following year *Flyer No. 3* made several longer flights – one lasting 38 minutes 3 seconds. In 1908, *Flyer No. 3* was modified to take a passenger and on 14 May, C. W. Furnas became the first passenger on a powered flight. Later in the year Wilbur demonstrated the latest *Flyer* – the model A – in France, and observers were astonished by its performance. After several long flights Wilbur remained airborne for 2 hours 20 minutes on 31 December 1908.

During the period 1903–1914, flying became a popular sport. Some enthusiasts designed, built and flew their own machines as a hobby. Many race-meetings and flying displays were held, and the public flocked to see the weird new machines perform. Records of speed, height and endurance were no sooner established than they were broken a short time later. In 1909 Louis Blériot made history by flying across the English Channel from France to England in less than 24 hours, and in 1911 the first officially sponsored airmail service was introduced, in England, between Hendon and Windsor.

Just when aircraft were emerging as a practical means of transport, the First World War broke out and they were hastily adapted for military purposes. The performance of aircraft and their engines improved rapidly under war-time conditions as each side tried to build better machines than their opponents. A typical flying speed of a scout or fighter in 1914 was 70 mph (113 kph) but by 1918 this speed had been doubled. Bombers too had improved and increased in size: when the war ended, large four-engined *biplanes* (planes having two wings, one above the other) were being produced which were capable of flying from France to bomb Berlin.

With the world at peace again, many pilots were looking for work and unwanted military aircraft were available at low prices: the stage was set for air travel to make a promising start. After some pioneering flights during the summer of 1919 the world's first daily commercial scheduled air service opened on 25 August between London and Paris. One aircraft to fly that day was a large converted bomber, the Handley Page 0/400 which could carry up to 11 passengers. A Handley Page Transport leaflet said:

'The seating accommodation is arranged as follows: After-cabin to seat six, forward-cabin to seat two passengers, and for those who prefer to travel in the open, two seats are arranged in front of the pilot. An observation platform with a sliding roof is fitted in the after-cabin.'

– a far cry from the modern jet airliner!

Above: *Orville Wright takes off, watched by his brother Wilbur, in 1903.*

Below: *The Handley Page 0/400 bomber turned airliner.*

TRAIL-BLAZERS

Although the first regular daily airline services started in August 1919 the fares were too high to attract many passengers, consequently there was not enough work for all the qualified pilots. Some of the more adventurous airmen turned their attention to conquering the continents and oceans of the world by air. The Atlantic Ocean was one of the first targets, especially since £10 000 had been offered by a British newspaper as a prize for the first non-stop flight. In May 1919 a United States Navy flying boat, the NC-4, did fly across the Atlantic but it landed at the Azores to refuel and so could not claim the prize. Two attempts to make a direct crossing were made in the same month but both ended in failure. Then on 14 June 1919, a converted Vickers Vimy bomber with a British crew took off from Newfoundland piloted by Captain John Alcock with Lieutenant Arthur Whitten Brown as navigator. Despite fog, rain, high winds and ice, they reached the coast of Ireland after flying for nearly 16 hours. Alcock and Brown crossed the Irish coast near the wireless station at Clifden only a little way off course, a remarkable feat of navigation after covering over 1890 miles. The first flight in the opposite direction was made a few weeks later by the British airship *R34* which then went on to make the return journey.

Many epic flights were made in these post-war years. The Australian brothers Keith and Ross Smith flew from Britain to Australia in 1919 and a year later Pierre Van Ryneveld and Quintin Brand flew to South Africa, both flights being made in Vickers Vimys. By 1924 the United States Army Air Service was ready to attempt a round-the-world flight, in short stages. Four specially

Above: *The Vickers Vimy bomber used by Alcock and Brown for the first non-stop trans-Atlantic flight in 1919.*

Epic flights across the world

Long ocean crossings continued to be a great challenge for the intrepid aviators. The Pacific was conquered in 1928 by the Australian Charles Kingsford-Smith and his crew in a three-engined Fokker F.VII monoplane called *Southern Cross*. In 1927, the American Charles Lindbergh flew solo from New York to Paris in 33.5 hours and became an overnight hero. Women pilots joined the men and keen competition developed to set up the fastest time from place to place. One of the most famous women pilots was Amy Johnson who flew from England to Australia in 1930. Delays on the latter stages of the flight ruled out a record, but she was the first woman to fly solo to Australia.

Right: *Amy Johnson and her Gipsy Moth.*

designed Douglas World Cruisers set out on 6 April and by 28 September two of the aircraft had completed the entire 25 439 mile journey. Another round-the-world flight made headline news in 1931, when an American pilot Wiley Post, and his Australian navigator Harold Gatty, completed the trip in 8 days 16 hours. Two years later, Wiley Post flew round the world again but this time he did it alone and became the first man to make this long journey solo.

Exploring new airways

Not all the pioneering flights were made to break records; some pilots turned their attention towards flights of discovery. A number of these flights were

Left: *Alan Cobham returns from Australia in 1926.*

Below left: *Explorer Roald Amundsen's airship* Norge *before its trans-polar flight in 1926.*

team led by Admiral Richard Byrd and Floyd Bennett, using a three-engined Fokker F.VII equipped with skis, and Amundsen, this time equipped with an airship flown by its Italian designer General Umberto Nobile. Byrd flew from Spitsbergen to the North Pole and back on 9 May 1926. Amundsen was one of the first to congratulate Byrd and, although disappointed, his party set out a day later in the airship. They reached the North Pole and continued to Alaska and so made the first flight across the polar region.

Many flights were made in the northern polar regions during the following years to investigate the possibilities of using this route for airliners. However, a survey of this route in 1933 showed that there were not sufficient airfield sites in these barren lands for the short-range airliners available at that time.

Flying in the Antarctic did not arouse as much interest as the Arctic because it was not on a possible air route, but, after several survey flights, Byrd flew over the South Pole on 28 November 1929.

also notable as first flights, but many were made mainly to survey the unknown areas, sometimes with a view to establishing air routes across them. Suitable sites for airfields or seaplane bases were sought, and observations on the prevailing weather conditions made.

One of the greatest pioneers of the airways was Sir Alan Cobham of Britain who set out to prove that an aircraft could fly long distances but still keep to a time schedule. He also surveyed the most suitable routes and possible airfields along them. Between 1924 and 1928 Cobham made round trip flights from Britain to Rangoon in Burma, Cape Town in South Africa, and Sydney in Australia, and also made a 19 320 mile flight around Africa. These flights had to be carefully planned: there were few airfields available, and gasoline and other supplies had to be transported to the proposed landing sites. Owing to this shortage of airfields, Cobham used a floatplane and a flying boat for his two longest flights because suitable stretches of water could usually be found for landing and taking off.

The vast unknown regions of the Arctic and Antarctic were a great challenge to airmen: the Swedish balloonist Andrée made an unsuccessful attempt to reach the North Pole by air in 1897 (*see page 3*) and in 1925 the great Norwegian explorer Roald Amundsen led an unsuccessful expedition which used two flying boats. Then, in 1926, two expeditions were preparing for an attempt at the same time. There was great rivalry between the American

Above: *American pioneer aviator Richard E. Byrd was the first person to fly over both Poles.*

POWER TO FLY

Once the Wright brothers, Wilbur and Orville, had learned how to fly a glider they had turned their attention to an engine and naturally had looked at the engines powering the recently invented motor car. The early car designers – and the early airship designers – had tried three sources of mechanical power: the steam engine, the electric motor and the gasoline engine. The gasoline engine was the most promising because it was lighter than the other two although it needed a fuel tank: the steam engine needed a boiler plus fuel storage, and the electric motor required batteries.

In 1885 a German engineer called Gottlieb Daimler and his friend Wilhelm Maybach built an engine which worked by exploding gasoline vapor inside a cylinder (hence the name *internal combustion engine*). The explosion moved a circular piston, which could slide to and fro inside the cylinder: this *reciprocating* movement was converted into rotation by means of a simple crank. Daimler fitted his engine into a primitive motorcycle. In the same year another German engineer, Karl Benz, working separately from Daimler, produced a three-wheeled car with a gasoline engine. The following year Daimler produced a four-wheeled car with a high-speed gasoline engine and this can be claimed to be the forerunner of modern automobile and aircraft engines.

Daimler's first engine had just one cylinder but most of the successful aero-engines had two or more. Many cylinder arrangements were tried but the one chosen by the Wright brothers had four, one behind the other or *in-line*. The four pistons moved up and down inside the cylinders and were connected to a crankshaft which rotated. The Wrights fitted two propellers to drive their aircraft through the air and these were driven by bicycle chains from the crankshaft. Most designers fitted a propeller to an extension of the crankshaft and mounted the engine in the nose of the aircraft. In-line engines sometimes had six or even eight cylinders but this arrangement resulted in a very long engine. A more compact engine could be built by arranging the cylinders around a short crankshaft. Since the cylinders radiated from the crankshaft this was called a *radial* engine and once again the propeller rotated with the crankshaft. A very unusual engine was produced in France just before the First World War. It looked like a radial engine but the whole engine rotated around the crankshaft which was fixed to the aircraft: the propeller was attached to the rotating engine. This arrangement created many problems but these were overcome and the *rotary* engine was a great success.

Gasoline engines became too hot if they were not cooled and, like most automobiles the in-line aero-engines were cooled by water flowing round the cylinders. This heated the water which

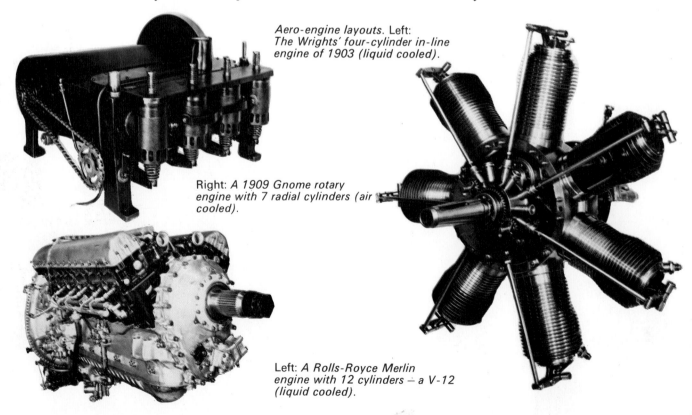

Aero-engine layouts. Left: *The Wrights' four-cylinder in-line engine of 1903 (liquid cooled).*

Right: *A 1909 Gnome rotary engine with 7 radial cylinders (air cooled).*

Left: *A Rolls-Royce Merlin engine with 12 cylinders – a V-12 (liquid cooled).*

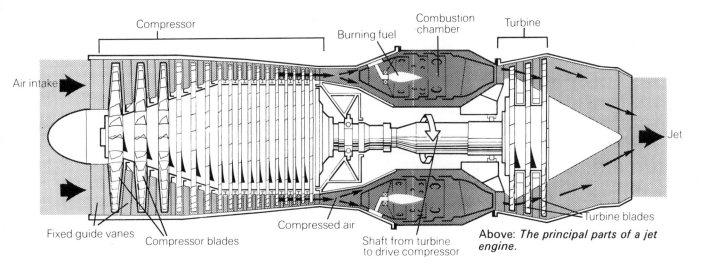

Above: *The principal parts of a jet engine.*

was then cooled in a radiator. Radial and rotary engines were usually cooled by air flowing over their cylinders and these air-cooled engines can be recognized by the fins on their cylinders which increase the effectiveness of the cooling air.

Between the two World Wars there was great rivalry between the world's leading aero-engine companies. In Britain, Rolls-Royce preferred in-line liquid-cooled engines while the Bristol company favored air-cooled radial engines. Both companies made some very successful engines, for example the Rolls-Royce Merlin in the Spitfire (introduced in 1936) and the Bristol Hercules which powered the later Wellington bombers (introduced in 1939). Two of the largest engine companies in the United States favored the radial layout: Wright produced their famous Whirlwind and later Cyclone engines, while Pratt and Whitney built the Wasp, Twin Wasp and Wasp Major. The Allison Division of General Motors concentrated on in-line liquid-cooled engines and these were fitted to several American fighter aircraft of the late 1930s. In Germany, another type of internal combustion engine was developed for aircraft use – the diesel. The Junkers factory built diesel engines for industrial use and they adapted their design for use in aircraft. The Junkers Jumo was probably the most successful diesel aero-engine and was used to power several German bombers and flying boats.

The jet engine

The invention of the jet engine revolutionized aircraft design and performance, yet the significance of the first jet aircraft passed almost unnoticed just before the Second World War. This was the German Heinkel He 178 which flew in August 1939, some two years before the first British jet engine was ready for flight testing. However, the Heinkel's engine was never developed for production, whereas many of the early British and American jet aircraft were powered by engines based on the British design developed by a Flying Instructor in the Royal Air Force called Frank Whittle (later to become Sir Frank Whittle).

Jet engines belong to a whole family of engines called *gas turbines*. In a gas turbine, air is drawn into the engine and compressed by an air compressor, then it passes to a combustion section into which fuel is sprayed and ignited. The fuel burns fiercely and produces hot gases at a high temperature and pressure. These gases expand and rush out from the rear of the engine through a pipe and nozzle as a jet which produces propulsive thrust. However, before emerging they pass through a turbine (a high-speed 'windmill'), and the revolving blades of the turbine turn a shaft which drives the compressor. If a larger turbine is fitted it can be used to drive a propeller in addition to the compressor, and this is called a *turbo-prop* or *prop-jet* engine.

The power of a jet engine can be increased by burning more fuel in the jet as it emerges from the engine into a special jet pipe, and this is called an after-burner or reheat. Although this is a relatively simple alteration it does use rather a lot of expensive fuel. A more efficient way to produce an engine of increased power is to design one with several compressors and turbines in stages. Each set of compressor-turbine units is called a spool: one of the first twin-spool engines was the Bristol Olympus which was later developed for use in Concorde. The next improvement in jet engine design was to allow some or all of the air from the first compressor to bypass the rest of the engine and emerge in the jet pipe. This *bypass* engine gives more power, yet uses less fuel. Finally a variation of the bypass, called a *turbo-fan* or *ducted-fan*, has been introduced in recent years to power the largest jet aircraft such as the Boeing 747 'Jumbo', the Lockheed TriStar and the Douglas DC-10. Even more air bypasses the engine and this is done by fitting a large compressor at the front of the engine which consists of a fan (a multi-bladed propeller) and is housed in a large tubular fairing (or duct) – hence the name *ducted-fan*. Such fan engines are not only more powerful and efficient but, since the bypass air surrounds and blankets the engine and jet, they are quieter too, which is an extremely important consideration nowadays.

13

HOW AIRCRAFT FLY

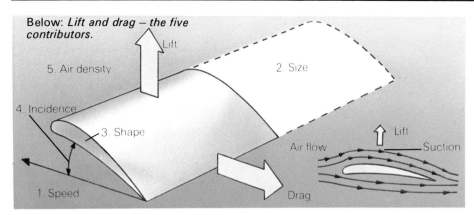

Below: *Lift and drag — the five contributors.*

A balloon lifts into the air because it is lighter than air, but an aircraft, which is heavier than air, must find some other way of producing *lift*. The wing of an aircraft moves quickly through the air and the flow of air over the wing produces an upwards force or lift. The study of air flowing around objects is called aerodynamics and one of the first aerodynamicists was the Ancient Greek philosopher Aristotle. He studied a projectile flying through the air and declared that in the wake of the projectile there was a region without air – in other words a vacuum. In this he was correct, but then his theory went astray for Aristotle suggested that the surrounding air rushing in to fill this vacuum pushed the projectile along. The opposite is in fact true, for this region of vacuum or suction behind the projectile tries to hold it back and this slows it down. The air resistance at the nose of the projectile, caused by a region of pressure, also retards the projectile and the combined effect is known as *drag*. All high-speed vehicles suffer from the effects of drag and these include aircraft, cars, motorcycles and railway trains but streamlining helps to reduce drag and so increase maximum speeds.

An aircraft in steady flight is subjected to a double tug-of-war. Lift from the wing pulls the aircraft upwards and balances the weight which acts downwards. Then in the horizontal direction, the thrust backwards from the engine causes a reaction which pushes the aircraft forwards, while the drag holds it back. Unfortunately the useful lifting force and the retarding drag depend on the same five things:

1. Speed – higher speeds produce more lift (and more drag).
2. Wing size – a larger wing or two wings give more lift (and more drag).
3. Wing shape – a thicker wing section gives more lift (and more drag).
4. Wing inclination or *incidence* – as the wing is inclined to the direction of flight, so it produces more lift (and much more drag).
5. Air density – the thin air at high altitudes results in less lift (and less drag).

Of these five factors, speed has the most marked effect because doubling the speed produces four times the lift and drag. But just when an aircraft needs all the lift it can get – at take-off – it is moving slowly. So can anything else be changed? The air density cannot be altered and it is not easy to increase the size of the wing very much. The angle of incidence can be increased by flying the aircraft in a nose-up attitude, but if this angle is increased too much then the flow of air becomes turbulent and the aircraft drops out of the sky in a stall. So the final item to be considered is the wing shape or *aerofoil section*. A flat wing section gives only a little lift, but a curved or *cambered* section increases the lift considerably. So the *flap* was devised to increase the camber of the lower surface of the wing.

A number of flap designs have been tried out over the years, ranging from a simple hinged flap or the split flap to the more complicated versions such as the slotted flap or the Fowler flap. The Fowler flap had a double advantage – it increased the camber and it moved out to increase the wing area. Lowering the flaps increases not only the lift but also the drag. This is a good thing during a landing when the aircraft is slowing down but not so good for take-off when the aircraft is accelerating. Conse-

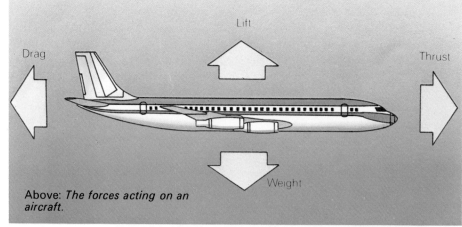

Above: *The forces acting on an aircraft.*

quently a pilot lowers the flaps only part of the way for take-off.

In the 1920s, a British aviation pioneer, Frederick Handley Page, developed another device to produce more lift from a wing at low speeds. It was called a *slat* and consisted of a mini-wing mounted along the *leading edge* leaving a *slot* between the slat and the wing. The Handley Page slat enabled a wing to fly at a much greater angle of incidence without stalling. The wings of modern jet airliners usually have flaps along part of the *trailing edge* plus a leading edge slat or drooping leading edge to increase the camber.

Controlling the aircraft's direction

Once it is airborne an aircraft must be under complete control. A ship is controlled by a movable control surface called a *rudder*, and most aircraft also have a rudder hinged to a vertical *fin*. The rudder controls movement to the left or right (*yawing*). But an aircraft has much more freedom of movement than a ship, consequently it needs more control surfaces. The up or down movement (*pitching*) is controlled by an *elevator* hinged to a horizontal *tailplane*. The elevator was sometimes called a horizontal rudder in the past and this was a very descriptive name. If an aircraft attempts to turn using just its rudder, it will slew around and slip sideways. To make a smooth turn it has to *bank* around with a wing dropping down and the other moving up (*rolling*). Rolling is controlled by *ailerons* which are hinged control surfaces on the trailing edge of each wing near the tip. When an aileron is moved down it increases the lift on that wing so by moving one side up and the other down the aircraft can be made to bank or roll.

On most modern aircraft the ailerons are controlled by turning a cut-down steering wheel on the *control column* in the cockpit. Turning the wheel clockwise makes the aircraft bank to the right. Some aircraft have a control column which moves sideways to control the ailerons. The elevators are controlled by moving the control column forwards or backwards. Pushing the control column forwards moves the elevator down and the aircraft nose dives. Finally, the rudder is controlled by foot pedals mounted on a hinged *rudder bar*. Moving the left foot forward and the right foot back swings the rudder to the left and the aircraft will yaw to the left but for a smooth banking turn the wheel would be turned anti-clockwise at the same time.

Above left: *An Airbus airliner with its trailing-edge flaps and leading-edge slats extended.* Left: *Shown in the normal flight position.*

AIR TRAVEL 1919-1959

Above: *An Armstrong-Whitworth Argosy airliner of Imperial Airways, 1926.* Inset: *The interior of an Argosy.*

When the world's first regular international air service opened in August 1919 between London and Paris the passengers were carried in converted bombers. The journey took anything from 2.5 to 5.5 hours depending on the headwinds, compared with 7 or 8 hours by train and steamer. Fares were very expensive, consequently only the rich could afford to fly. Film stars were frequent passengers; not only did flying save time but also it attracted the press photographers! From November 1919, mail was carried officially on the London to Paris route, and because people were prepared to pay extra for the speed of air-mail, it provided a very useful source of income to struggling airlines.

By the mid 1920s most of the European capitals were linked by air and the first aircraft designed to carry passengers were replacing the converted bombers. One of the first true airliners was the Armstrong Whitworth Argosy of 1926, a large biplane with three engines, used by the British airline Imperial Airways on their European routes. The Argosy could carry 20 passengers in reasonable comfort at a steady 93 mph (150 kph). However, its range was only about 300 miles so it had to land many times on a long journey: nevertheless the airways expanded and by 1929 Britain was linked with India.

Many famous airlines were flying regular services by 1930 – and most of them are still operating today. Within the United States there were four very large airlines: American, Eastern, United and TWA (now Trans World Airlines), while Pan American concentrated on international services. In Europe a number of airlines had become well established with both inter-city flights and long-distance services: several European countries still had overseas empires. The Netherland's airline KLM (Royal Dutch Airlines) flew to the Dutch East Indies (now Indonesia) and Belgium's Sabena flew to the Belgian Congo (now Zaire). Several small French airlines operated services to French possessions in Africa and the Far East, and these joined forces in 1933 to form Air France. The German airline Deutsche Lufthansa was the largest in Europe: they flew to most European cities and pioneered the use of flying boats to South America. Most of these European airlines used aircraft built in their own country; for example, Deutsche Lufthansa had an all-German fleet of 220 aircraft.

In 1931 a new airliner, capable of carrying 38 passengers, had been introduced by Imperial Airways but its speed and range were little better than the Argosy. However, the new Handley Page HP42 gained a great reputation for comfort, reliability and safety. In the following year a regular service was started between London and Cape Town, South Africa, but a passenger had to be prepared for a long and arduous journey. There were 33 stages and it took 11 days (assuming there were no delays) to cover nearly 7800 miles – an overall average of under 30 mph (50 kph). The fastest alternative was to travel all the way by surface transport and this would have taken 17 days.

Below: Helena, *a Handley Page HP42 airliner.*

The revolutionary flying boats

In 1936, Imperial Airways received the new Short 'Empire' flying boats which revolutionized their long-distance routes. The first of these four-engined monoplanes could carry 17 passengers in luxury, plus two tons of mail or freight, and cruise at 160 mph (260 kph). But most important, their increased range reduced the number of refuelling stops by half. Consequently, when the Empire flying boats were introduced on the service to Cape Town, in 1937, they cut the time from 11 to 6.5 days. The following year this was further reduced to 4.5 days. Several nations went on to build impressive flying boats, which were used to pioneer Atlantic and Pacific crossings until war in 1939.

Below: *A Short Empire flying boat.*

All-metal airliners

The airlines flying across America needed fast, long-range airliners which could operate from the many airfields which had been built during the 1930s. In 1936 an airliner called the Douglas DC-3 was in service for the first time for American Airlines, on their trans-continental route. It was an all-metal aircraft of advanced design, powered by two powerful engines which enabled it to carry 21 passengers at 170 mph (275 kph) over distances in excess of 480 miles.

When the Second World War started in 1939, airline services were severely reduced and many of the factories that had produced airliners turned over to military transport aircraft – especially in the United States. By the end of the war a new generation of four-engined long-range aircraft had emerged. Two of the most successful were the Douglas DC-4 and its rival, the Lockheed Constellation. The Constellation could carry 47 passengers over a distance of 1500 miles and cruise at 280 mph (450 kph) – a speed which only ten years earlier would have way outstripped even the fastest fighter planes.

Above: *The Lockheed L-049 Constellation began service in 1946, and was used by the British Overseas Airways Corporation among others.*

Left: *The Douglas DC-3 airliner first flew in 1936, and some are still in service.*

When the war ended, some airlines started operations using surplus military transports converted to carry passengers, but the large national airlines were more interested in the new four-engined airliners which included enlarged or 'stretched' versions of existing aircraft. The most popular trio were the Boeing Stratocruiser, the Douglas DC-6 and a stretched Lockheed Constellation. These each carried 70–80 passengers at about 310 mph (500 kph) over a distance of 3000 miles – still using conventional petrol engines.

For a period of 10 years from 1947 the Douglas and the Lockheed companies competed with each other, introducing a new variant of the four-engined airliner every few years. Boeings retired from this competition and concentrated their efforts on designing a jet-engined airliner. By the end of this 10-year period, airliners with the traditional petrol engine, such as the Douglas DC-7C and the Lockheed Starliner, were carrying 100 passengers at 340 mph (550 kph). But this was not good enough, and even their long range of around 6000 miles did not save them when jet-engined airliners made their first appearance in the late 1940s.

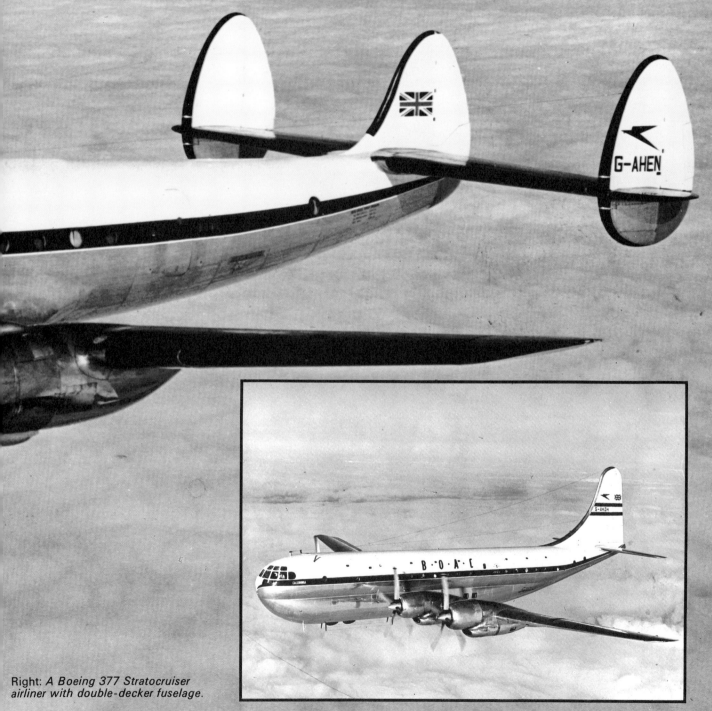

Right: *A Boeing 377 Stratocruiser airliner with double-decker fuselage.*

THE JET AGE

Above: *The first Vickers Viscount.*

Jet aircraft were used during the Second World War but in its early days the jet engine was not very suitable for airliners because its fuel consumption was too high. Most of an airliner's load-carrying ability would have been taken up with fuel instead of fare-paying passengers or *payload*. In July 1948, the world's first completely jet-driven commercial aircraft flew from London to Paris in 34 minutes but this new airliner could not carry enough payload to be a financial success. It was an experimental Vickers Viking fitted with two Rolls-Royce Nene jet engines in place of its usual Bristol Hercules piston engines: the Viking was a 27-seater airliner developed from the faithful old Vickers Wellington bomber of the Second World War.

However, in that same month Vickers-Armstrongs produced another new British airliner which was to make a much greater impact on the airways of the world – this was the Viscount. The Nene-Viking and the Viscount were both fitted with gas-turbine engines, but the former used pure-jets in contrast to the Viscount which carried four Rolls-Royce Dart turbo-prop engines. One great advantage of the turbo-prop was that it used much less fuel than the pure-jet. Yet despite its low fuel costs, high cruising speeds and a very quiet passenger cabin, the Viscount did not sell. In fact it was very nearly scrapped. Then British European Airways showed interest, and during the 1950s the Viscount emerged as one of Britain's most successful airliners.

After the Viscount, Vickers-Armstrongs produced the Vanguard. This was another turbo-prop airliner but considerably larger, carrying 139 passengers, yet it was extremely economical to operate. On all but the very short hops it could fly with two-thirds of its seats empty and still make a profit. Despite this advantage only two airlines

Below: *A Sud-Aviation Caravelle.*

bought the Vanguard, for the fare-paying passengers were choosing to fly in a rival pure-jet airliner. The French twin-jet Caravelle needed to fly almost full in order to pay its way because it was powered by pure-jets which had a high fuel consumption. The designers of the Caravelle mounted its engines on the side of the *fuselage* behind the passenger cabin, which resulted in a very low noise level in the cabin. During the 1960s the attraction of jet-travel plus the quiet cabin, ensured that the Caravelle was nearly always full – so despite its fuel consumption it was a great success.

Another turbo-prop airliner, the Lockheed L.188 Electra, was built in America but was not successful. The most successful of all the western turbo-prop transports was the Dutch Fokker F.27 Friendship, which first entered service, with West Coast Airlines of America in 1958, accommodating 48 passengers. In Europe, it was used by Aer Lingus, and it is still in production.

The Boeing 707, a new American jet airliner introduced in 1958, followed the layout of the old piston-engined airliners with four engines mounted on the wings in contrast to the Comet which had its four engines buried inside the wing to give a more streamlined shape. A wing free of engines is said to be a 'cleaner' design and of course the Caravelle with its two engines attached to the sides of the fuselage had a clean wing. This arrangement was also used very successfully by the BAC One-Eleven. De Havilland followed the Comet with the Trident, a three-engined airliner which had one engine on each side of the rear fuselage and one in the center with its intake in the base of the fin. Vickers built the four-engined VC10 with two engines on each side of the rear fuselage. In all these European designs the wings were free of protruding engines, but in America the Boeing 707 was followed by the Douglas DC-8 and Convair's 880 and 990 and these aircraft all had four engines on the wings.

Airliners of the sixties

Boeing and Douglas went on to produce whole families of airliners. Boeings have always favored wing-mounted engines except for the three-engined 727 which has the same layout as the Hawker Siddeley Trident. The Boeing 727 made history in 1973 when the thousandth model rolled off the production line: no civil aircraft in the world had ever achieved this figure and orders continued to arrive. The Douglas DC-9 with two fuselage-mounted engines also proved to be very popular with sales figures around the 1000 mark. These jet airliners are typical of the 1960s but in the 1970s two new dimensions in airliner design appeared – the 'Jumbo' jet and the supersonic airliner. Supersonic air travel became a possibility in December 1968 when the Russian Tupolev Tu-144 made its first flight, to be followed two months later by the Anglo-French Concorde. However the Concorde won the race to carry passengers on regular services when flights to Rio de Janeiro in Brazil and Bahrain started in January 1976.

The large airliners of the 1960s were designed to carry about 180 passengers and these were later 'stretched' by lengthening their fuselages. For instance the original Douglas DC-8s carried a maximum of 173 passengers whereas the final version could carry 257 but the seating arrangements remained the same with three seats on either side of a central gangway. The airlines wanted even larger airliners so the designers had to turn their attention to the width of the fuselage and the 'wide-body' jets emerged. The first of these was the Boeing 747 Jumbo which entered airline service in 1970 and could carry up to 490 passengers sitting

Above: *The four jets of a Vickers Super VC10.*

10 abreast. Like the earlier 707, the 747 had four engines mounted on the wings. Rival wide-bodied airliners soon followed but the Douglas DC-10 and the Lockheed TriStar both had three engines, one on each wing and one in the lower part of the fin. In Europe the Airbus Industrie A-300 was produced, by Germany, France, the Netherlands, Spain and Britain in collaboration, for shorter routes and it used just two wing-mounted engines.

First jet passenger service

The world's first regular passenger service to be flown by a jet airliner was the British Overseas Airways Corporation's service to Johannesburg, South Africa. On 2 May 1952 a de Havilland Comet left London with 36 passengers and reached its destination in less than 24 hours having flown 6493 miles. The Comet was powered by four de Havilland Ghost jet engines. Unfortunately on two occasions Comets crashed due to a structural weakness and the airliner had to be redesigned. In 1958 the new Comet 4 began services across the Atlantic.

Right: *The first de Havilland Comet.*

A MODERN AIRLINER

One of the most common sights at any large international airport is the original 'Jumbo' jet – the Boeing 747. When it first appeared in the early 1970s it caused quite a sensation because it was so much larger than the previous generation of jet airliners, such as the widely used Boeing 707. The 747's height is perhaps its most noticeably different dimension, for the highest point of its fuselage is almost 37 feet above the ground compared with 20 feet for the 707. An average two-storeyed house stands about 20 feet high, but the 747 has three 'storeys'. At the lowest level there is a freight hold, then comes the main passenger cabin and finally a spiral staircase leads up to a smaller passenger cabin and the cockpit or *flight deck*. The interior of the main passenger cabin, like the rest of the aircraft, is huge, with seats for 10 people side-by-side instead of the usual six across. There are two gangways and the seats are grouped three-four-three. The number of passengers carried varies, depending on the airline's requirements, but two popular layouts are: 66 first-class plus 308 economy-class seats, or 490 economy-class seats. To look after the passengers, about 15 stewards or stewardesses are carried, so, with a flight crew of two pilots and a flight engineer, the total number of people on board could exceed 500. Some layouts include a lounge and bar on the upper deck, and film shows are an optional extra.

Sitting in the comfortable seats, high above the clouds, passengers may not realize that the air outside is so 'thin' that they would not survive for very long if the cabin were not *pressurized*. Mountaineers and the crews of high-flying military aircraft breathe oxygen through a face mask: this would not be a practical or popular solution for airline passengers! An alternative solution is to compress the thin air using air compressors and pump it into the sealed *pressure cabin*, bringing the pressure back to a comfortable level. Airliners fly at these high altitudes to save fuel, because the thin air offers less air resistance or drag. An added advantage is the smoother air, because most of the storms and severe gusts of wind occur at lower altitudes.

The pilots of a 747 are seated some 30 feet above the ground and 99 feet forward of the aircraft's main wheels, and these distances can take some getting used to when maneuvering along a winding taxi track before take-off. The 747 has an impressive array of wheels – the main landing gear consists of 16 wheels mounted as four-wheel bogies on four separate legs. When the *undercarriage* designers set to work on the 747, they were able to fit a simple twin-wheel nose gear but they could not use the usual eight-wheel main gear (two four-wheel bogies) be-

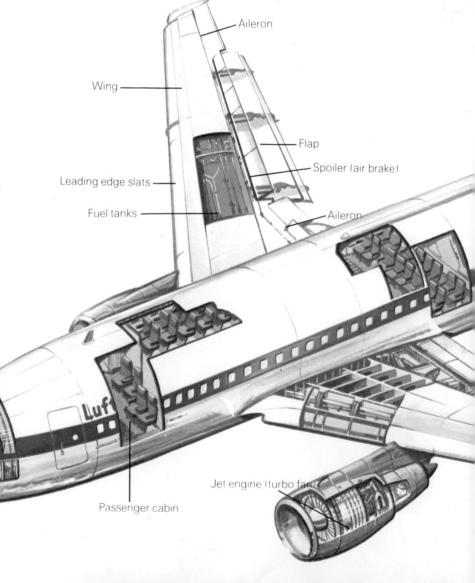

cause the weight of the 747 was more than twice the weight of existing airliners. The runways of the world's airports just could not cope with double the landing impact load under each wheel, so the designers had to double the number of wheels instead.

Flight controls

Once the Jumbo lifts into the air it is no more difficult to fly than any other jet airliner – in fact many pilots claim that it is easier than some smaller aircraft. At first sight the cockpit appears to be a mass of instrument dials, electrical switches and levers. In fact these are highly organized and many are standard for all modern airliners. In front of each pilot are the instruments which give information about the flight of the aircraft – its speed, height, direction and so on. Between the pilots, and positioned so that both have easy access, are the engine controls. There are four levers which are the throttles controlling the power from each of the four engines, and above these are rows of instruments, in fours, giving details of each engine's performance. Above the windshield are a host of switches controlling the lighting, engine starting and electrical services. The flight engineer's panel gives details of the fuel situation, the cabin pressurization, the movable control surfaces, and the electrical power supply.

From the cockpit, and hidden away behind the furnishing panels or below the floor, run a multitude of pipes, wires and cables which link the vital systems. The pilot's control column has to be connected to the elevators and ailerons; the rudder pedals to the rudder. In the early days simple cables were used, but the heavily loaded control surfaces need power-assisted systems – just as many cars today have hydraulic brakes and power-assisted steering. A *hydraulic* system works like a bicycle pump in reverse. If air is blown into a bicycle pump, its handle will move out: similarly, if high-pressure hydraulic fluid is pumped into a *jack*, a rod emerges and this can be linked to a moving part. The hydraulic system is used to move control surfaces, retract the undercarriage, open and close doors, steer the aircraft on the ground and work the brakes. The system is triggered by mechanical links to the pilot's controls, but this linkage is progressively becoming electrical rather than mechanical.

All aircraft need electricity and this is supplied from generators driven by the engines. This power is used to supply cockpit instruments, radio sets, radar equipment, navigational aids, lighting and heating to keep the food warm in the galleys. The fuel system also needs electricity to drive its pumps, but it is a complex system in its own right. Fuel has to be pumped to each engine at the correct rate, and this is made more complicated by the fact that any one airliner has many fuel tanks. As the fuel is used, it is important to keep the balance of the aircraft correct; for instance, taking fuel only from the rear of the aircraft would lead to it becoming nose heavy. All these systems are vital to the safety of an airliner so should one fail, a second and even a third has to be available to take over in an emergency. Safety is all important in the design of a modern airliner.

Left: The Airbus A300B wide-bodied airliner built by a consortium of companies from Britain, France and West Germany. This illustration shows an Airbus in the colours of the German airline Lufthansa.

FASTER THAN SOUND

During the later years of the Second World War, pilots of the new, high speed fighter aircraft ran into trouble when their aircraft went into a dive. The pilots experienced severe buffeting and the trim (or balance) of their aircraft changed suddenly. Sometimes these strange effects were so serious that the aircraft broke up in the air or the pilot completely lost control. When the newspapers heard about these accidents, the reporters blamed the mysterious 'sound barrier', for the aircraft must have been approaching the speed of sound. At lower speeds the disturbances in the air, caused by the aircraft's movement, can escape as waves moving at the speed of sound, but as that speed is approached, the disturbances ahead of the aircraft cannot escape fast enough. The building up of air waves results in a sudden increase in the drag on the aircraft and the formation of shock waves. Indeed, the complete airflow around the aircraft changes. In the case of the Second World War fighter aircraft, these changes were critical, but specially designed aircraft were able to cope with the problems of supersonic flight. However, one difficulty remained: the shock waves, which could be heard on the ground as a *sonic boom*. This loud boom follows the aircraft all the time it is flying at supersonic speeds and is a serious problem for supersonic aircraft flying over inhabited areas.

The speed of sound varies at different heights, so it is difficult to keep track of what its speed is. Therefore, instead of measuring the aircraft's speed in miles per hour, it is measured by comparison with the speed of sound – which is called *Mach 1*. (The name honors an Austrian scientist Ernst Mach who studied the behavior of sound waves.) An aircraft flying at twice the speed of sound is said to be flying at Mach 2 and the speed is indicated on a machmeter. Despite the predictions by scientists and aerodynamicists about what could happen when an aircraft reached Mach 1, no one could be really certain. Usually, advanced new designs for aircraft were tried out by making a model which was tested by blowing air over it in a *wind tunnel* but the shock waves produced at Mach 1 made this kind of test unreliable.

Although the first generation of jet airliners such as the de Havilland Comet and the Boeing 707 were subsonic, they flew at speeds approaching

Below: *Air flow shown in a wind tunnel.*

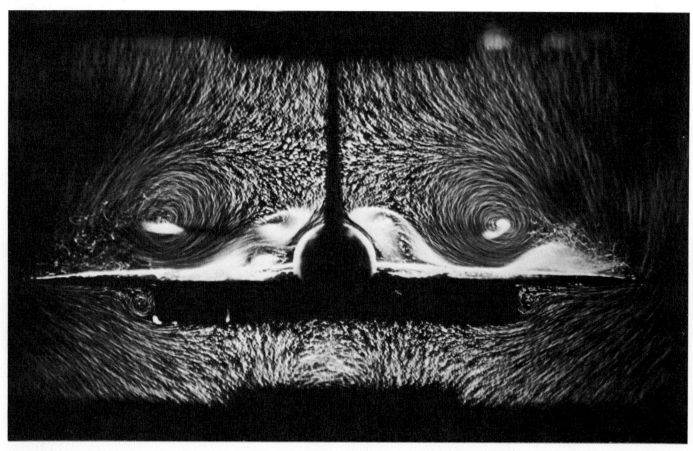

The first supersonic aircraft

The first man to exceed the speed of sound in level flight was Captain Charles ('Chuck') E. Yeager of the United States Air Force. He named his Bell X-1 rocket-powered aircraft 'Glamorous Glennis' after his wife but officially it was number 6062. On 14 October 1947 Chuck Yeager achieved a speed of Mach 1.015 or 670 mph (1078 kph) and proved that there was no sound 'barrier'. The cautious preparations leading up to this flight ensured that Yeager could control the X-1 as it went supersonic. The research program into supersonic flight continued and improved versions of the X-1 were designed. In June 1954 Yeager reached a speed of Mach 2.42 in the X-1A. But these aircraft were rocket-powered flying laboratories and it was some time before supersonic aircraft were ready for military service. Supersonic airliners were even further away in the future.

A supersonic Bell X-1 research aircraft.

Mach 1 and this presented problems to the designers. For instance, as air flows over a wing the speed of the air increases – an increase which could take it up to the critical Mach 1. It was found that this effect could be reduced by sweeping the wings backwards. The Boeing 747, with a maximum cruising speed equivalent to Mach 0.9, is slightly faster than the 707 – so its wings have just a little more sweep back.

Right: *Swept-back wings reduce air resistance as the speed of sound (Mach 1) is approached. Here, three airliners and their cruising speeds are compared. The smallest is the De Havilland Comet (Mach 0.74). The largest is the Boeing 747 (Mach 0.90), and Concorde (Mach 2.04) is half-way between the other two in size.*

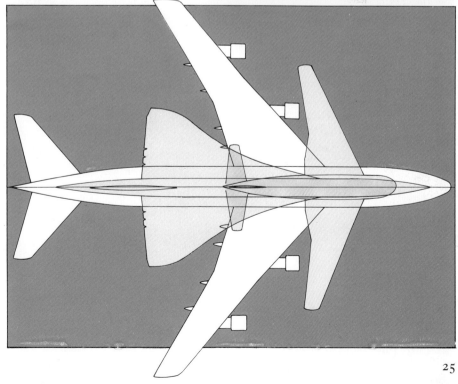

Concorde

Around 1960, several countries which already had supersonic military aircraft were considering designs for a supersonic airliner, including Britain, France, Russia and the United States. In 1962, British and French aviation experts combined to share the design and construction work for a project which was named Concorde. Many shapes were investigated: the wings had to be swept back very sharply and this brought the wing tip very close to the tailplane. The designers decided to combine wing and tail into a large triangular shape. This was called a *delta wing* – named after the Greek letter 'capital delta' which is a triangle. Military aircraft had used this delta shape already but the Concorde designers settled for a very long thin triangle – a *slender delta*. Two research aircraft were built to test this shape – one at low speeds and the other at supersonic speeds. The final Concorde wing had a slightly curved leading edge – a shape known as an *ogee*.

To drive an airliner carrying 100 passengers through the air at supersonic speed requires engines of great power to overcome the immense drag. In the case of Concorde this power is supplied by four Rolls-Royce (Bristol)/Snecma Olympus engines. One of these engines supplies 37 972 pounds of thrust during take-off, which is just about the same power as the four engines mounted in the de Havilland Comet airliner. Keeping the passengers and the aircraft cool is another major problem. At very high speeds the friction between the air and the airliner's skin generates enough heat to raise the temperature of the skin above the boiling point of water. To keep the inside of the passenger cabin at a comfortable temperature, the skin has to be lagged and a refrigerating system installed. However, the effect of this heat on the metal airframe is more serious because the strength of the aluminium alloys used in aircraft construction drops very rapidly as the temperature rises. The designers of Concorde had to allow for this reduced strength and make allowances for the expansion of the various parts as they became hot.

Concorde began to carry fare-paying passengers in January 1976. Timetables become very confusing when traveling at supersonic speeds, because the airliner moves around the Earth faster than the Earth moves around the Sun. New York time is five hours behind London time, so if a supersonic airliner leaves London at 8.00 am and takes four hours for the journey, it arrives in New York at 12.00 noon (London time) which is 7.00 am in New York. Strange though it sounds, the passenger arrives one hour before leaving – just in time for a second breakfast! Of course, flying in the opposite direction adds time to the journey.

Right: *An Avro Vulcan bomber used to test Concorde's Olympus engine.*

Below: *Concorde being serviced at Heathrow airport, London.*

BUILDING AN AIRLINER

Modern airliners such as the Concorde and the Boeing 747 Jumbo look very solid, but their structural shell has to be as light as possible otherwise they would never leave the ground. Weight has been a problem for aircraft designers since the days of the pioneers who ventured into the air in flimsy gliders. Otto Lilienthal and Percy Pilcher based their gliders on soaring birds, so they built bird-like skeletons of willow or bamboo covered with a strong fabric. To make the framework strong enough for flight, an array of external bracing wires had to be added. But some other pioneers decided that the box kite layout could be adapted to make a glider and so a different shape emerged with two wings, one above the other. Octave Chanute and the Wright brothers made such biplane gliders, and of course the Wrights went on to build the first successful powered machine – their 1903 'Flyer'. Although these biplanes were a very different shape from the gliders of Lilienthal and Pilcher, the materials used in their construction were the same – wood, wire and fabric. This kind of structure, nicknamed 'stick and string', remained in use throughout the First World War and even lasted until the Second World War.

Over the years, improvements were made to increase the strength of aircraft structures, while keeping the weight to a minimum. One change was to use thin plywood for the skin instead of fabric. Plywood was glued to the framework, thus increasing the overall strength because the plywood could carry some of the loading whereas fabric could not. Because the plywood skin was carrying some of the stresses and strains, this arrangement was called *stressed-skin* construction. Other designers introduced thin steel tubes in place of the wooden framework, while others used a mixture of wood and metal – but in both cases the fabric and wires remained.

Aluminium is much lighter than steel but it is very weak and consequently not suitable for a highly loaded aircraft structure. However, small quantities of other metals such as copper and magnesium can be added to aluminium to produce an *alloy* which is very much stronger than pure aluminium. During the 1930s a new aluminium alloy called *duralumin* became widely used and this was almost as strong as steel, although much lighter. Some aircraft were built using a duralumin framework covered with fabric, but the great step forward came when stressed-skin structures were designed in this aluminium alloy. One of the first aircraft to use stressed-skin construction was the Douglas DC-3 airliner of 1935. Its smooth lines and robust structure made it one of the most successful aircraft of all times – indeed several examples are still flying. The DC-3 has been called 'the first modern airliner' and the stressed-skin structures of the Douglas DC-10 and the Boeing 747 have many features in common, but of course the fantastic increase in aircraft size has necessitated certain changes to the major structural members.

Supersonic airliners, such as Concorde, need a very special structure because of the problems imposed by flying at very high speeds. The basic shape of the aircraft has to be different, so the designers cannot base the structure on any previous designs. Supersonic speeds result in very high aerodynamic loadings (intense air pressures) which impose high stresses on the structure; the speed also results in an increase in the skin temperature due to the rubbing action which takes place between the skin and the air. The skin becomes so hot that aluminium alloys lose their strength. For the parts which bear the brunt of this heating effect, alternative materials have to be found: the two most widely used are stainless steel and a relatively new material called *titanium* which is lighter than steel but very strong.

Below: *Aircraft structures old and new.*
Left: *A 'stick and string' framework.*
Right: *Modern all-metal stressed-skin construction.*

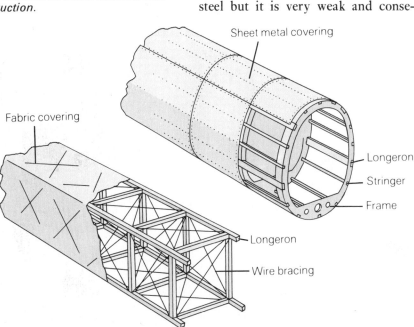

Design and testing

The design of an airliner involves many hundreds of experts: aerodynamicists, draughtsmen, stress engineers, systems experts and weights engineers. Once the drawings have been produced planners decide how the parts are to be made; the special tools and equipment are prepared and then production begins. Until recent years, most aircraft structures consisted of parts cut and shaped from sheets of aluminium alloy and then joined together by rivets or nuts and bolts. Modern structures are often carved in one piece from a large block of metal by machines which appear to work themselves: actually they are controlled by a computer which has been carefully programmed to perform every operation automatically. Once the parts have been manufactured they are assembled and the shell of the airliner takes shape, but there are still many months of work ahead fitting the engines, undercarriage, instruments and lengths of pipes and wires.

Before the airliner is allowed to fly, a long program of testing has to be carried out. Two complete airframes may be built and tested to find the point of destruction: one is tested under the most severe single loading the aircraft might experience and the other under the repeated loading of a *fatigue* test. For instance, a wing flexes up and down, and this relatively low loading, applied frequently, can start a minute crack which will grow until the wing breaks. This is called a *fatigue failure* and it was this type of structural weakness which caused the early de Havilland Comets to crash. There are many other ground tests designed to check all the moving components and the systems. Hundreds of hours are then spent on test flying: first to ensure that the airliner can fly safely, then to prove that it can carry out all the tasks for which it was designed. If it passes all these tests it will then be allowed to carry passengers – but not before.

Below: *The wing of an Airbus airliner in a test-rig at a German factory.*

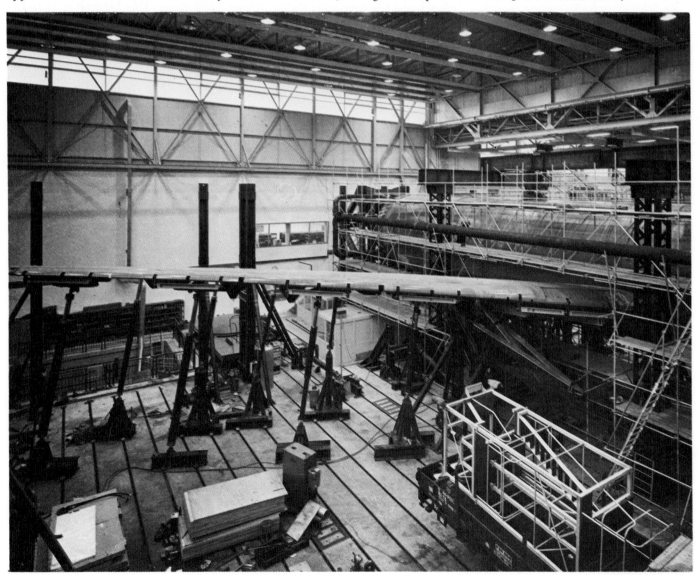

FLYING AN AIRLINER

Airliners have taken over from the grand ocean liners on almost all the passenger routes which link the continents of the world and this take-over includes usage of many of the old nautical traditions and titles. An airliner is commanded by a Captain who is the senior pilot, and the co-pilot is called the First Officer. Their cockpit is referred to as the *flight deck* while the passengers are housed in a *cabin*. Other members of the crew include Stewards and Stewardesses who look after the passengers and serve food from the *galleys*. On long-distance flights the crew may also include an Engineer, a Navigating Officer or a Radio Officer. Because modern radio equipment is extremely reliable and simple to operate, the pilots usually cope without a Radio Officer. Navigating Officers, too,

Above: *The airport groundcrew in bright clothing guide each aircraft to its parking place.*

Below: *Air traffic controllers following an aircraft's progress on a radar screen.*

are becoming redundant as electronic equipment and computers take over. The Flight Engineer supervises a large instrument panel and his main responsibility is to ensure that the engines are running smoothly and using the correct amount of fuel. He must also ensure that the fuel is taken from the correct tanks otherwise the balance of the airliner could be upset.

Long before an airliner departs, work begins on the preparations for the flight. The Operations Department of the airline ensures that airliners have a complete crew – a task which needs careful planning, especially on the long distance routes such as Britain to Australia. The airliner and passengers will make the complete journey but the crew will change because the number of hours they spend in the air is strictly limited. A weary pilot might make a mistake and there is no room for any error when lives are at stake. Another section of the Operations Department will probably prepare a *flight plan*. This gives details of the aircraft's weight including passengers, cargo and fuel; it also sets out the route for the flight with all the speeds, heights, estimated arrival times and many other factors which have to be taken into account.

The Captain and his crew report for duty about an hour before the airliner is due to depart. The Captain checks the flight plan and consults the latest weather reports, which may alter it – for instance, strong headwinds may force him to carry more fuel. Once he is satisfied with the flight plan the Captain briefs his crew and the plan is sent to the *Air Traffic Control* center. If this authority is satisfied with the plan, then details are transmitted by teleprinter to controllers along the route and to the destination airport.

Take-off and landing

The crew board the airliner and, while the cabin staff look after the passengers, the Captain and First Officer (or co-pilot) carry out the pre-flight cockpit checks. A long list of items to be checked is worked through systematically. Once all the passengers are aboard, the doors are closed, then the servicing vehicles are moved safely out of the way and the pilot is given permission to start the engines. This comes by radio from the Ground Controller, at Air Traffic Control, who also gives clearance for the aircraft to taxi out to the runway. After a final check on the engines and flying controls the pilot requests permission from Air Traffic Control to take off. When this permission is received, the engine power is increased and the airliner surges forward. At a pre-determined speed, the pilot eases the control column back and first the nose lifts, then the aircraft becomes airborne.

Once clear of the airport the pilot climbs to his cruising height along one of the *airways*, which are like main roads of the air (*see pages 32–33*). At this point the pilot can switch control to the *automatic pilot* or *autopilot* which does the routine work of flying the airliner along its set course. The autopilot is controlled by *gyroscopes*. (A gyroscope is a rotating wheel so mounted that it is sensitive to any change in the direction of its axis.) Should a gust of wind divert the aircraft from its course these gyroscopes 'feel' this movement and make automatic corrections to the aircraft's control surfaces. The first automatic pilot was built as long ago as 1912 by Lawrence Sperry in the United States and was fitted to a Curtiss floatplane (*see page 44*).

As the airliner flies along the airway its progress is reported to the Air Traffic Controllers on the ground. In busy areas this progress will be followed by ground radar operators. Incidentally, the airliner also carries radar equipment which scans the sky ahead and warns the pilot if the aircraft is flying into a storm or some obstruction.

As the airliner nears its destination airport, the pilot comes under the control of the Airport Control Tower where radar operators follow every aircraft, and control their movements by sending instructions over the radio. Sometimes a line builds up of aircraft waiting to land, and the Controller has to order the aircraft to fly in circles until their turn comes. Each aircraft is allocated a different height in this *stack* and the Controller tells them when they may descend to a lower level. Once a pilot reaches the bottom of the stack it is his turn to land. He can be guided to the end of the runway either by the well-tried *Instrument Landing System* (ILS) equipment or the Approach Controller can 'talk him down' using radar. In each of these cases the pilot takes over for the actual touch-down, although in recent years some airliners have been equipped for fully automatic landings.

An airliner arriving at a busy airport may have to line up in a stack.

Holding pattern approach route

One aircraft at each level

Levels separated by 1000 feet

Control tower

Runway

Radio beacon

AIRWAYS

In the early days of aviation, pilots found their way by observing the land below and map-reading with the help of a compass. Railway lines and rivers were a great help – in fact some railway stations had their names painted on the roof especially to help pilots. When oceans had to be crossed there were no such landmarks, consequently the pilot had to rely on his compass to guide him in the right direction. But a compass course needed to be checked, so the airmen copied the sailors and navigated by the Sun and stars. They used a *sextant* to calculate latitude and longitude, but the calculations took a long time, with the result that a fast aircraft would have moved on by the time they were completed!

Between the two World Wars, radio began to be used to assist the navigator. An aircraft transmitted a radio signal which a ground station picked up on a *direction-finding aerial* – the aerial rotated until it pointed in the direction of the aircraft. If a second station picked up the signal, the bearings of the two stations enabled the aircraft navigator to *fix* his position. During the Second World War, new radio aids to navigation were invented to help bombers find their targets over enemy territory – where there were no helpful direction-finding ground stations. The very successful British systems were called *Gee* and *Loran*. From these the *Decca Navigator* was developed which, using radio signals from the ground, showed an aircraft's position on a moving map in the aircraft. In the United States another system was developed called *VOR/DME* (Very-high-frequency Omni-directional Radio-range/Distance Measuring Equipment). This could be described as a lighthouse sending out radio signals instead of light. An instrument in the aircraft picked up these signals and worked out a fix. Both the Decca system and VOR/DME needed ground stations, so for long distance flights over the sea some other system was needed. First came *Doppler* which used radar mounted in the aircraft to measure its track and speed, then the *Inertial Navigation System* which is also used to navigate in outer space. This very complicated system uses gyroscopes, acceleration-measuring instruments and computers.

All these aids to navigation helped to prevent airliners from getting lost, but on some busy routes there was another hazard – the possibility of a mid-air collision. It became clear that airliners had to be controlled in some way – especially over Europe and the United

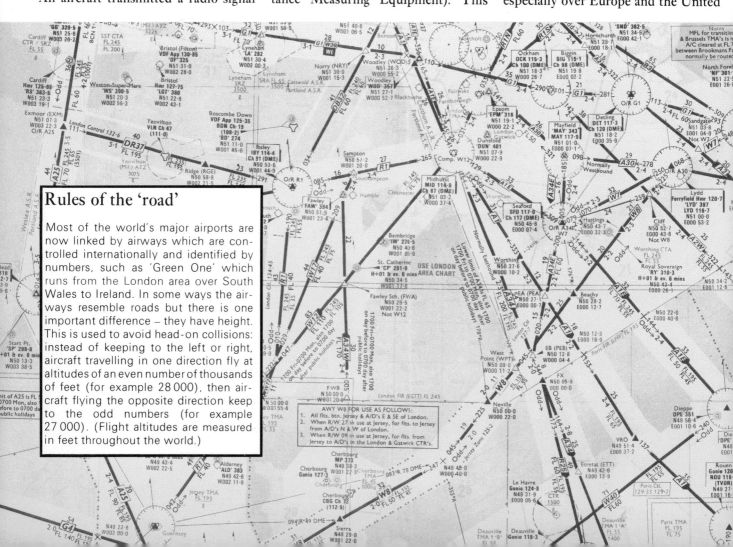

Rules of the 'road'

Most of the world's major airports are now linked by airways which are controlled internationally and identified by numbers, such as 'Green One' which runs from the London area over South Wales to Ireland. In some ways the airways resemble roads but there is one important difference – they have height. This is used to avoid head-on collisions: instead of keeping to the left or right, aircraft travelling in one direction fly at altitudes of an even number of thousands of feet (for example 28 000), then aircraft flying the opposite direction keep to the odd numbers (for example 27 000). (Flight altitudes are measured in feet throughout the world.)

States. *Airways* were introduced as the 'highways of the air' to simplify the problem of navigating and to ensure that airliners did not collide with each other. A radio beam was transmitted from a *radio-range* station and aimed along an airway. Following this beam was so simple that the pilots could navigate without the aid of a navigator. They just listened to the station's signal on their headphones and when they heard a steady note the aircraft was on course. To obtain a fix the pilot tuned in to a *fan-marker* beacon which transmitted a signal upwards. A receiver in the aircraft picked up this signal only when the aircraft was passing over the beacon. The old radio beacons have now been superseded by the modern systems of radio navigation – primarily VOR/DME.

Controlling air traffic

Just as any road system needs traffic police to enforce the safety rules of the Highway Code, so the airway system has to be controlled. This is obviously a problem since traffic lights or a policeman cannot be positioned at the intersection of two airways. All the controlling has to be done from the ground by Air Traffic Controllers using radio and radar. This is organized in stages depending on conditions, for instance the pilot of an airliner on a trans-Atlantic flight finds that mid-Atlantic traffic is very different from that on the approach to London's Heathrow Airport or New York's John F. Kennedy International. Large areas, such as the North Atlantic, are divided up into *Flight Information Regions* and each region has one Control Center. The controllers tell the pilot at what height and speed he should fly, and then they ensure that no other aircraft fly at that height unless they are a safe distance ahead or behind.

As an airliner crosses the Atlantic and nears Britain, it is picked up by the radar of the London Air Traffic Control Center at West Drayton, Middlesex. The controllers of this busy center use a very advanced computer system called *Mediator* which is linked to three radar sites. The radar scanners track each aircraft in the area and the computer predicts their flight path. To do this the computer has to store a host of information such as details of each aircraft's performance, the weather forecast, and airway rules and regulations. Using Mediator the controllers guide each airliner to within 12 miles of Heathrow. At this point the airliner is handed over to the Approach Controllers who are based in a darkened room crowded with radar screens. Their job is to maintain a steady flow of aircraft to or from the runways. The final stages of a landing are controlled from the glass-house on the top of the control tower by the Air Controllers. After landing, the Ground Controllers direct the airliner to its parking area.

Below: *Part of a British Airways Radio Navigation Chart covering from southern England to northern Europe.*

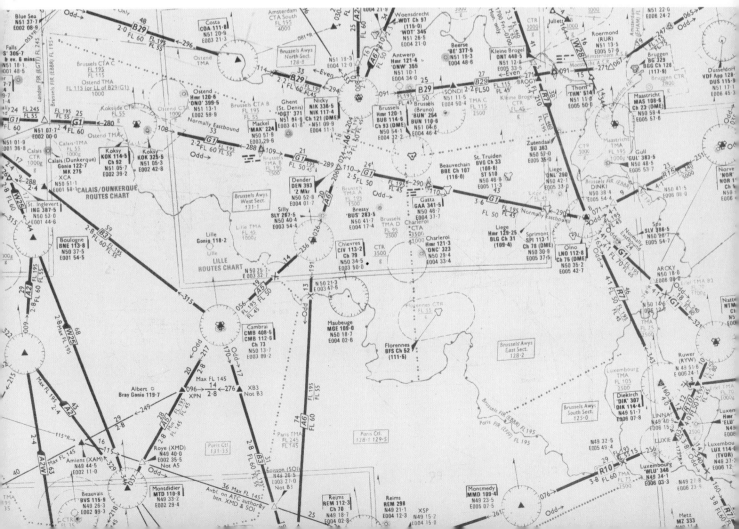

AIRPORTS

In the early days of aviation, aircraft could fly from any large field and these fields became known as flight grounds or flying-grounds. Gradually, aircraft hangars and buildings for the passengers were added and the name aerodrome was used. In the 1920s some of the larger civil aerodromes were called *airports* for the first time.

Today's airports have come a long way from the early days of those at Newark and Croydon when passengers were measured in thousands. John F. Kennedy Airport, New York and London Heathrow handle many millions of passengers every year and the airports never stop expanding. In fact a large modern airport can be compared to a town with a working population of 50 000 or more. There are shops, banks, restaurants, hotels, buses, taxis, cars and parking lots. In addition there are all the workers who ensure that everything runs smoothly – telephone operators, receptionists, police, firemen, electricians and cleaners. All these services and people could be found in any town, but at the airport there are also many staff working for the airlines. Some of these look after passengers on the ground and in the air, while others service the aircraft between flights and overhaul them at regular intervals.

When an airliner lands after a flight it is immediately surrounded by a fleet of vehicles of all shapes and sizes. One of the first to arrive is a mobile power unit which supplies electricity to the aircraft, since its own generators do not function when it is stationary. Then there is the air-conditioning truck to keep the passenger cabin at a comfortable temperature. Sometimes fuel is brought to the aircraft in tankers but at many modern airports fuel is piped to the servicing areas and a hose connects the aircraft to a *hydrant*. Fuel is pumped aboard at very high speed – it would fill an automobile's gas tank in a fraction of a second. Oil and water are both brought to the aircraft in small tankers. Engineers give the aircraft a routine between-flights inspection to make sure that there are no loose parts or fuel leaks. Other vehicles are busy looking after the passengers' requirements – transporting baggage, removing rubbish, bringing fresh food and drink and emptying the toilets.

Every few weeks the aircraft is taken out of service for a minor overhaul, and there are graded checks up to a major overhaul which involves stripping down the aircraft to inspect every component. Naturally this takes a long time but it only has to be done every few years. By careful planning the airlines try to arrange for these overhauls to be carried out in the winter months when they are not so busy with holiday traffic. Before any aircraft in the world can fly, it must have a *Certificate of Airworthiness* ('C of A') and this is issued by a government department. In the USA, inspectors of the Federal Aviation Administration ensure that aircraft meet the high standards of safety demanded before issuing a 'C of A'.

Below: *The central area of London's Heathrow airport, seen from an airliner's window.*

The first airports

In 1920, London's first real airport was opened at Croydon and like all airports at that time it had grass runways. A great step forward was made in 1928 when New York's first airport was opened at Newark for it had a hard-surfaced runway. Croydon remained a grass airfield until it was closed in 1959. During the Second World War many of the fighter bases were just grass airfields, but the heavy bombers and transport aircraft needed hard-surfaced runways and some of these airfields were converted to airports after the war.

Below: *Croydon Airport in the 1920s.*

Below: *Modern airport layouts ensure that passengers can board aircraft without being exposed to the weather.*

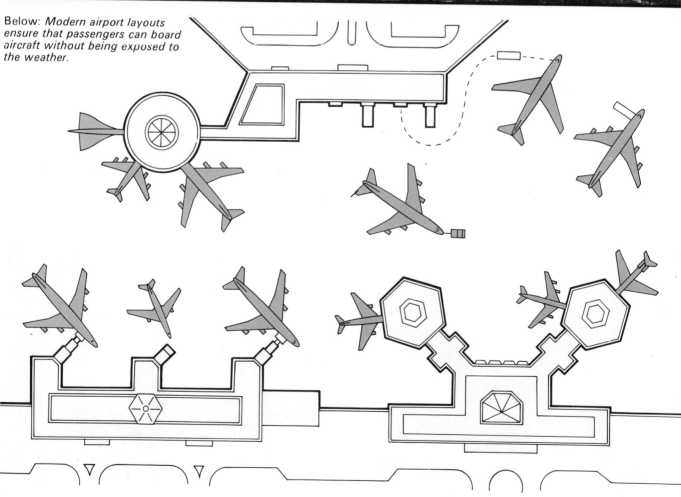

Getting passengers to the aircraft

When passengers arrive at an airport, the first thing they do is *check-in*. This is a simple operation for passengers, consisting of a few formalities such as having their baggage weighed. The airline staff then have to ensure that the luggage is loaded on to the correct flight and that the aircraft is not overloaded. The flight number is called over the loudspeaker and after a short wait, passengers are shepherded through the security search and, on international flights, their passports are checked. Passengers then board the waiting aircraft where a stewardess welcomes them aboard and inspects their tickets. When all the passengers are aboard and the doors closed, the engines are started under the supervision of the ground crew and firemen. The airplane is then ready for the flight to begin.

Building a new airport presents many problems for the surveyors, engineers and meteorologists looking for a perfect site. If the airport is to serve a city it must be near the city center, but on the other hand the noise of its aircraft should not disturb the residents. Runways have to be built on solid ground to withstand the landing impact of large airliners and they also cover a large area; consequently the only available site is usually far from the center of the city. The ends of the runways must be free of dangerous obstructions such as tall buildings and hills. Weather conditions, too, have to be studied to ensure that the area is not prone to mist and fog, but there should be an alternative airport somewhere near – just in case.

One of the major problems for designers of large airports is how to get the passengers from the terminal building to the parked airliner without making them walk an excessive distance. Many airports have *fingers* radiating from the terminal building and an adjustable covered ramp from the finger to the doorway of the airliner. As more aircraft use the airport so the fingers become longer and longer. One solution is to install moving walk-ways along the fingers, but completely different solutions have also been tried out. At some airports, buses are used to transport passengers from the terminal building to parked airliners, but a Jumbo needs a fleet of buses. This arrangement was taken a stage further at Washington DC's Dulles International Airport where huge mobile lounges were introduced. In contrast Los Angeles International was built with a number of satellite terminals – in other words there are several mini airports on one site.

Above: *Passengers disembarking from a Boeing 707 in southern Peru.*

Below: *One way to save passengers from a long walk – a mobile lounge in Mexico.*

WORLD AIRLINES

There are about 500 airlines in the world and this figure does not include companies which fly only small aircraft. Of course some of the 500 have very small fleets, for example the French airline Air Fret has just one Boeing 707 and a staff of 35 employees – but this small airline has been flying since 1964. At the other extreme is the Russian airline Aeroflot, with about 500 000 workers and an aircraft fleet of thousands. It is a little unfair to compare Aeroflot with other airlines because it is really a government department which is responsible for virtually all civilian aviation in the USSR. Aeroflot carries passengers and cargo like any other airline but it also operates agricultural and ambulance aircraft and carries out fishery, ice and fire patrols.

The largest privately owned airline in the world is United Airlines of the United States, with a fleet of 330 aircraft and a staff of 54 000. The history of United Airlines goes back to 1931 when four pioneer airlines joined together, one of these, Varney Air Lines, having been founded in 1926. Today, United Airlines operates a large network of routes inside the United States and a few overseas services. United's foremost rivals within the United States are Eastern Air Lines, who introduced the *shuttle* service idea (a service where the passengers don't need to book in advance, and may pay for tickets on the aircraft), and American Airlines who, over the years, have sponsored many new designs such as the Douglas DC-3 of 1936. The two major airlines of the United States flying on international routes are Pan American World Airways and Trans World Airlines. 'Pan Am' operates a round-the-world service and flies to most capital cities in the world.

In Europe the largest airlines are British Airways, Air France and Lufthansa German Airlines. Of these, British Airways was the largest though is now having to reduce its staff and fleet of aircraft. An unusual feature of this fleet was that it consisted of over 20 different types of aircraft ranging from Concorde to small helicopters, due to the merger of the fleets of two national airlines in 1972. British Airways had the largest network of routes in the world covering some 480 000 miles and serving 77 countries.

Air France also has a world-wide network of routes which includes several supersonic services flown by Concorde. British Airways and Air France started their Concorde operations simultaneously in January 1976, and now these include flights to North America, South America, the Middle East and the Far East. Air France also operates a night airmail service within France which is noted for its very high standard of reliability and regularity. Many of the

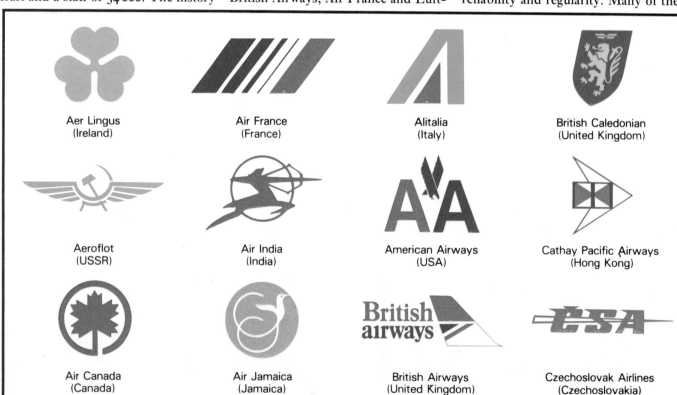

Aer Lingus (Ireland)
Air France (France)
Alitalia (Italy)
British Caledonian (United Kingdom)
Aeroflot (USSR)
Air India (India)
American Airways (USA)
Cathay Pacific Airways (Hong Kong)
Air Canada (Canada)
Air Jamaica (Jamaica)
British Airways (United Kingdom)
Czechoslovak Airlines (Czechoslovakia)

European airlines, founded in the 1920s, are still flying extensive services – often to their former colonies. KLM (Royal Dutch Airlines) began operations in 1919 and claims over 60 years of continuous operation. The Belgian airline Sabena, and Lufthansa German Airlines were founded in the 1920s, but in contrast the Scandinavian Airlines System which pioneered flights to America via the North Pole was founded as recently as 1946.

Although many of the largest airlines are based in Europe and the United States, there are some very important companies in other parts of the world. Air Canada has over 20 000 employees and the Brazilian airline Varig is only a little smaller. QANTAS of Australia claims to be the second oldest existing airline – just a few months younger than KLM. But QANTAS set another record in 1979 when it sold its last Boeing 707, because this meant the airline was flying just one type of airliner – the Boeing 747 Jumbo. Of course a fleet of 20 Jumbos can carry a very large number of passengers and the servicing problems are simplified.

Air India was founded after the Second World War, in 1946, and has a fleet of 19 aircraft with over 12 000 employees. Even more recently, Singapore Airlines started in 1972 and now has a fleet of 38 aircraft.

International organizations

Because there are so many airlines operating world-wide services an international body was set up to ensure that all the airlines and the Air Traffic Controllers have a standard way of working. The International Civil Aviation Organization (ICAO) was founded as an offshoot of the United Nations and today about 150 countries are members. It is vitally important that all aircraft observe the same rules of the air and procedures, especially when approaching an airport. Amongst items covered by ICAO are such things as maps, radio and navigational aids, aircraft inspection and licenses for personnel.

There is a second international organization involved in air travel but this one represents the airlines themselves and is called the International Air Transport Association (IATA). The work of IATA which affects the traveler most directly is reflected in the airline ticket, for international fares are agreed between IATA members. Making a reservation seems to be a very simple operation, but behind the scenes there is a large staff using a computer system to check reservations and make sure aircraft are not overfilled. Sometimes a journey may involve two or more flights on different airlines but this is no problem – one ticket is sufficient. This interchangeable ticket is very convenient for the air-traveler but requires much work behind the scenes to ensure that the airlines which carry the passenger get the money. The IATA *clearing house* staff work out how much money is due to each member airline. Many other financial, commercial and technical problems are solved by the joint efforts of the airlines through IATA.

Below: *Airline insignia from around the world.*

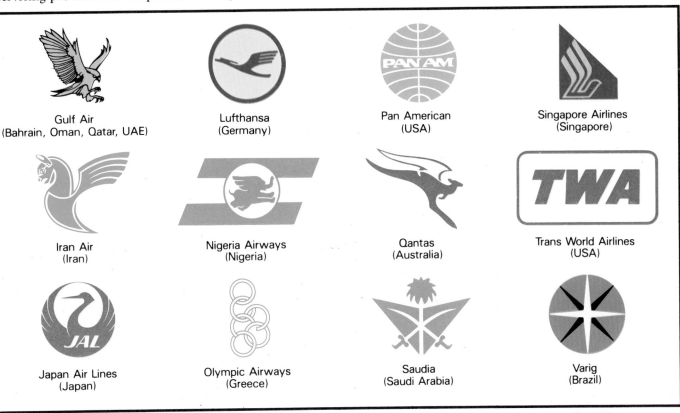

FREIGHT AIRCRAFT

During the years between the two World Wars, a typical load of air freight consisted of passengers' baggage and a few bags of airmail letters. These were usually carried in the small freight compartment of a scheduled passenger airliner. Contracts to carry airmail were very much sought after by the early airlines as they struggled to survive. During the mid 1920s the Post Office in the United States offered lucrative contracts to carry airmail and special mail planes were designed. One of the best of these was the Boeing Model 40 biplane which could also carry two passengers.

Aircraft carrying only freight played a vital part in supplying the armed forces during the Second World War. The majority of these freighters were airliners, such as the Douglas DC-3, modified by fitting a stronger floor and larger doors. After the war many old freighters were bought by airlines and a new industry of air-freighting was born. Sending freight by air is generally more expensive than by rail, road or sea but it does reach its destination more quickly. For instance speed is essential for the transportation of mail, newspapers, perishable items – such as flowers – and urgently needed spare parts for broken machinery. An unusual freight operation started in July 1948 when Silver City Airways opened a car-ferry service between England and France. Instead of a ferry-boat, they used freight aircraft – the specially designed Bristol 170 Freighter. Two cars and their passengers were carried each trip, which lasted less than half an hour.

Cargo aircraft

Today air freight (or air cargo, as it is now often known) has grown into an industry in its own right employing many thousands of people and transporting anything from day-old chicks to space rockets. The aircraft in use can be divided into three groups: old passenger aircraft modified to carry cargo, airliners redesigned as freighters, and specially designed freight aircraft. (Military freighters are usually called transport aircraft.)

The first group contains some straightforward conversions such as the Vickers Vanguard which was fitted with a larger door and a stronger floor to become the Merchantman. Other conversions produced some more unusual aircraft – none more so than the Guppy. In 1963 the United States space program needed a large aircraft to transport sections of the Apollo-Saturn V moon rocket to Cape Kennedy: no existing transport aircraft could do the job. An old Boeing Stratocruiser airliner was rebuilt with a huge bubble on the upper part of its fuselage, so that, with the nose hinged to open sideways a rocket up to 25 feet in diameter could be carried. This aircraft was called a Guppy because of its resemblance to the fish of that name. An improved Super Guppy with turbo-prop engines was also built and this aircraft has been used by the builders of the European Airbus to transport large sections of aircraft to the assembly factory.

Manufacturers of most modern airliners usually offer a choice of layouts for any particular aircraft. For example, Boeings offer three very different layouts for their 747 Jumbo; a conventional airliner carrying passengers, a convertible aircraft which can switch between passengers and cargo, or a specialized cargo aircraft. The cargo version has a nose section which hinges upwards to reveal a large hold capable of carrying an incredible 198 000

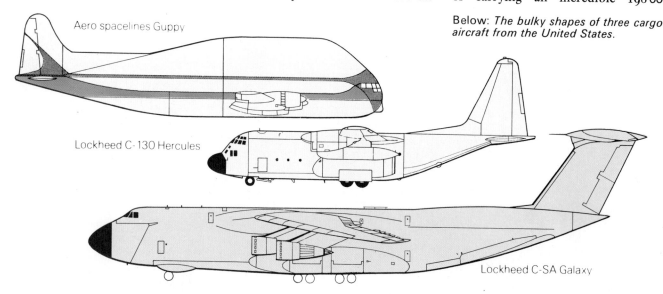

Below: *The bulky shapes of three cargo aircraft from the United States.*

Aero spacelines Guppy

Lockheed C-130 Hercules

Lockheed C-5A Galaxy

40

The Berlin Air Lift

Freight aircraft can sometimes fly in supplies when no other form of transport is available – such as in the northern areas of Canada where the 'bush' pilot carries vital supplies to isolated communities. In June 1948 a whole city became isolated: the Russians refused to allow supplies for West Berlin to travel by either road or rail through East Germany, thus cutting West Berlin off from the rest of West Germany. Britain and America decided to fly-in the vital supplies and the 'Berlin Air Lift' began. Not only food had to be carried by air, but also bulky items such as coal, industrial materials, gasoline, in fact everything a city of two million people required to live and work. It was a mammoth undertaking which called for all available freight-carrying aircraft. At times aircraft were landing at the rate of one every three minutes. After almost a year the Russians relented and reopened the roads and railways to West Berlin: the Air Lift had beaten the blockade. It was a very expensive operation and in normal circumstances it would not be practicable to fly such a bulky substance as coal, but it did demonstrate the effectiveness of air transport.

Below: *American and British aircraft at Gatow Airport, Berlin, for the Air Lift.*

pounds of cargo inside.

Of the specially designed freight aircraft flying today, the Lockheed C-130 Hercules is probably the most widely used, since it is in service with a number of commercial cargo airlines and many air forces. It first entered service in 1956 and by 1978, 1500 had been delivered. The Hercules has four turbo-prop engines and can carry over 40 000 pounds of cargo or 92 fully-equipped soldiers, but it is dwarfed by two later Lockheed military transports – the C-141 StarLifter and the C-5A Galaxy. The Galaxy can carry 345 soldiers or a load of tanks and trucks.

During recent years new systems of loading cargo into an aircraft have been introduced to reduce the amount of time an aircraft wastes on the ground. Instead of loading individual items into the aircraft, one at a time, they are packed onto a movable platform (or *pallet*) or loaded into a *container*. This loading is carried out in a special cargo shed, which means it can be done before the aircraft arrives. A loaded pallet is usually covered with a net to secure the cargo. The pallets and containers are then weighed and moved onto a truck which takes them to the waiting aircraft. They are lifted aboard by a forklift truck, loaded into position and locked to the floor of the aircraft. Pallets and containers are made to international standard dimensions so they can be carried by any aircraft, or transferred to land or sea transport.

An important aspect of air cargo which does not lend itself to containerization is the transportation of live animals. Race horses are frequently flown to distant race courses: sheep and cattle are also frequent travelers by air. Most airports have special facilities for dealing with this rather unusual cargo – which needs regular food and water. Some 'passengers' need more attention than others, for instance when dolphins are carried they have to be carried in special foam-lined slings and sprayed with water at regular intervals! It has been claimed by one airline that every fourth 'passenger' carried is an animal.

FLYING BOATS

Once an aircraft is in the air, its wheels or other type of undercarriage are just dead weight and a source of drag, but during take-off and landing the undercarriage is vital. The Wright brothers devised an unusual undercarriage: their 'Flyers' did not have wheels – they had two skids. European aviators preferred wheels, for example the monoplane designed and flown by Louis Blériot on the first cross-Channel flight in 1909 had two wheels at the front and one at the rear. During the 1930s, undercarriages were greatly improved, first by the introduction of more efficient springing, then by making the wheels fold up inside the wing or fuselage. These *retractable* undercarriages reduced the drag during flight, but in general were heavier than fixed wheels.

As airliners grew in size and weight they needed longer runways to become airborne. They also needed firmer runways to withstand the impact of landings. Finding suitable airfields was a problem, but there was an alternative and this was first demonstrated in 1910 when a French aircraft with a very different kind of undercarriage made its first flight. Henri Fabre's peculiar looking flying machine was fitted with floats and took off from water. Many seaplanes followed and perhaps the most spectacular were the racing *floatplanes* flown in the famous Schneider Trophy races in the late 1920s. In 1931 Britain took the Trophy outright having won it three times in succession, and the winning Supermarine S.6B went on to break the world air-speed record at 407 mph (655 kph). Larger seaplanes were often designed as *flying boats*. In these, the fuselage or *hull* floated on the water and small, steadying floats were fitted at the wing tips. In the early days of long-distance flying, seaplanes were very convenient because a suitable stretch of water could usually be found for landing or taking-off, whereas aerodromes were few and far between.

The first aircraft to fly across the Atlantic Ocean was a flying boat, which had to stop and refuel during the journey. The United States Navy Curtiss NC-4 flying boat alighted on the sea at the Azores Islands which are about two-thirds of the way between Newfoundland, Canada, and Lisbon, Portugal. The idea of alighting on the sea to refuel during long flights was used by the German airline Deutsche Lufthansa in the 1930s. Their Dornier Wal (Whale) flying boats operated a mail service to South America. They could not reach their destination non-stop, so they alighted on the sea near an anchored depot ship which lifted them aboard with a crane. After being refuelled, the flying boat was catapulted into the air to continue its flight. This catapult launch saved a considerable amount of fuel, because a take-off from the sea required a long run to become airborne.

Long-distance flying boats

During the 1930s, flying boats reigned supreme on the long-distance air routes of the world. Imperial Airways started many of its services with biplane flying boats such as the Short Calcutta and later introduced the very popular Short 'C' class Empire flying boat which was a four-engined monoplane. In 1931 the United States airline Pan American Airways introduced the Sikorsky S-40 *amphibian* which could operate from land or water. Then, in the mid 1930s, Pan American extended its services with the larger Sikorsky S-42, the Martin 130 and the Boeing 314. These four-engined flying boats were christened 'Clippers' and they pioneered routes across the Pacific and Atlantic Oceans.

Many other countries built flying boats and seaplanes, but the Second World War ended most passenger

Below and right: *The largest aircraft ever constructed, the Hercules flying boat, which had a wing span of 320 feet and was intended to carry 700 passengers. It was built by American millionaire industrialist, aviator and film producer Howard Hughes, and made its only flight in 1947.*

Above: *Howard Hughes built his gigantic Hercules with a wooden framework, which resulted in its nickname the* Spruce Goose.

Below: *The United States Navy Curtiss NC-4 seaplane at Lisbon, Portugal after flying from Newfoundland via the Azores. This was the first aircraft to cross the Atlantic, taking off from Canada on 16 May and landing 11 days later.*

services, and flying boats became widely used for patrolling the seas in search of enemy shipping. Aircraft such as the Short Sunderland of Britain and the Consolidated PBY-5 Catalina of the United States were outstanding at this.

When peace returned flying boats were superseded by the new land-based airliners, but many smaller flying boats, floatplanes and amphibians were built for island services and for use in the Canadian bush where lakes are plentiful. Will the large flying boat return? Supporters of the flying boat have always maintained that very large aircraft should operate from water, and not from long and expensive runways. In 1929 the Germans built the Dornier Do.X with a wing span of 158 feet; in 1935 the French built the Latécoère 521 with a wing span of 163 feet, then in 1947 an American millionaire, Howard Hughes, built his Hercules with a span of 320 feet. The whole air-

craft was built of wood and was intended to carry 700 passengers. This still holds the record as the largest aircraft ever built, but it made just one flight. None of these extra-large flying boats, nor the British Saunders-Roe Princess, with a span of 220 feet, were produced in large numbers. In the post-war years there were plenty of runways and it became increasingly difficult to keep long stretches of water clear of ships, boats and floating debris – a log floating just below the surface is a major hazard for a flying boat. Flying boats in the future would also need bases and few of the pre-war bases survive – consequently the future of the flying boat does not look promising.

Right: *A British Supermarine S.6B floatplane in 1931.*

Pickaback aircraft

One solution to the problem of getting a very heavily loaded aircraft into the air was tried out in Britain during 1938. Major R. H. Mayo suggested a pickaback arrangement with a large aircraft lifting a smaller one into the air. The two aircraft were built by Short Brothers of Rochester: the lower one was a large flying boat called 'Maia' while the upper one was a floatplane called 'Mercury'. The composite aircraft took off, and when a suitable height had been reached, Mercury was released to fly its heavy load of mail to a distant destination while Maia returned to base. Promising trials came to an end with the outbreak of the Second World War.

Below: *The Short-Mayo composite aircraft.*

HELICOPTERS

Above: *Early helicopters and autogiros.*
(Cornu helicopter, France, 1907; Cierva C8L MK II autogiro, Great Britain, 1928; Flettner Fl 282 Kolibri helicopter, Germany, 1940; Sikorsky VS-300 helicopter, United States, 1942 version)

Below: *Rescue at sea – a Sea King helicopter and the lifeboat from Cromer, England.*

Not very many years ago the sight of a helicopter flying overhead was a rare event, but today these *rotary-wing* aircraft hardly rate a second glance. The development of the helicopter has really taken place since the end of the Second World War, although its actual history goes back much further. In the 15th century Leonardo da Vinci sketched a helicopter which had a large spiral 'screw' to lift it vertically into the air. Indeed the name helicopter comes from the Greek words *helix* (a screw) and *pteron* (a wing). Like so many of Leonardo da Vinci's inventions this one was never built – it was just an idea. Another idea for a helicopter was published in 1843 by the British inventor Sir George Cayley and this had four lifting *rotors* (large propellers) but again it was never constructed. Many weird and wonderful designs followed during the 19th century, but none of these made even a short flight – some merely shook themselves to pieces!

In 1907 two French helicopters did actually leave the ground. One designed by Louis Breguet and Professor Richet reached a height of 5 feet but had to be steadied by men holding on to it. The second, built by Paul Cornu, made a free flight lasting just 20 seconds and reached a height of about 1 foot – later it rose to nearly 6 feet!

In the 1920s a great step forward was made when the Spanish engineer, Juan de la Cierva, invented the *autogiro*. Both the autogiro and the helicopter obtain their lifting power from a large, horizontal rotor with two or more blades, each acting as a rotating wing. The basic difference between the two flying machines lies in the method of driving this rotor. It is driven by an engine in a helicopter, but rotates freely in an autogiro. In flight, the rotor of an autogiro is blown around due to the forward speed of the aircraft. A piston engine and propeller are usually fitted to provide this forward speed. When an autogiro takes off it requires a short run to start the large rotor turning, and of course a head wind helps. It can climb almost vertically, fly very slowly and land vertically.

The success of the autogiro encouraged helicopter designers and by 1937 the Germans had produced an experimental helicopter. It could not lift a heavy load, but could remain in the air for over an hour and it was reasonably controllable. Controlling a helicopter in flight was a major problem for the early designers. A few years later in America, the first usable helicopter was designed and flown by the Russian-born Igor Sikorsky. His VS-300 set the pattern for the majority of helicopters built since that date.

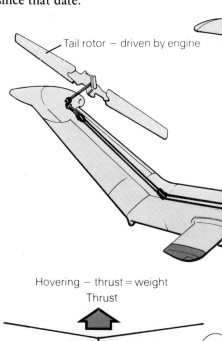

Tail rotor – driven by engine

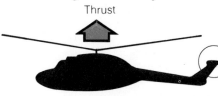

Hovering – thrust = weight
Thrust

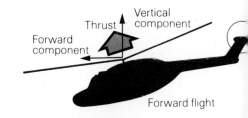

Thrust / Vertical component / Forward component / Forward flight

Controlling a helicopter

Sikorsky's helicopter had a large lifting rotor and a propeller at the tail facing sideways: both were driven by the same piston engine. Without the propeller (or a small rotor) the fuselage would have rotated in the opposite direction to the main rotor. Several Sikorsky designs were hurried into production and over 400 had been built by the end of the war. The Sikorsky layout of rotors has remained popular since those early days but several other layouts have emerged. For example, two main rotors rotating in opposite directions eliminate the need for a tail rotor.

When a helicopter is hovering, the upward thrust from the rotating rotor blades just balances the weight of the machine. If the pilot wants the helicopter to move upwards, this thrust must be increased. The pilot operates a lever in the cockpit which is called the *collective pitch lever* and this increases the pitch (angle of twist) of all the rotor blades. If each rotating blade is compared with an aircraft wing, then this operation increases the angle of incidence of each 'wing' and hence the lift (*see page 14*). More lift requires more power from the engine, so the collective pitch lever incorporates a twist grip throttle control (like that of a motorcycle) which means the pilot can control the two operations with one hand.

The pilot also needs to control his direction of flight – forwards, backwards or even sideways – and once again the pitch of the rotor blades is used. The blades trace out a large disc as they rotate, so if the pitch angle is increased in the rear half of this disc and decreased in the forward half, more lift will be produced over the rear half than the front half. This results in the disc (and the helicopter) tilting into a nose-down attitude. The thrust from the rotor is thus divided into an upwards component and a component pulling the machine forwards. The pilot controls the direction of the rotor's thrust by means of a control lever called the *cyclic pitch control* which is usually mounted between his legs – like the control column of a light aircraft. To control the direction in which he is heading, the pilot operates rudder pedals which are linked to the sideways-facing tail rotor.

Almost every day helicopters are in the news, usually because they have rescued someone in distress. But in addition to their rescue missions helicopters are used as a means of transport, as lifeboats, cranes, observation posts, ambulances, trucks, agricultural implements and as weapons of war. The secret of the helicopter's success lies in its ability to hover and operate from any small open space. But helicopters do have some disadvantages, such as their noise. They are also expensive, both to buy and to operate, a fact which is largely due to their rotors and the complicated mechanism controlling them.

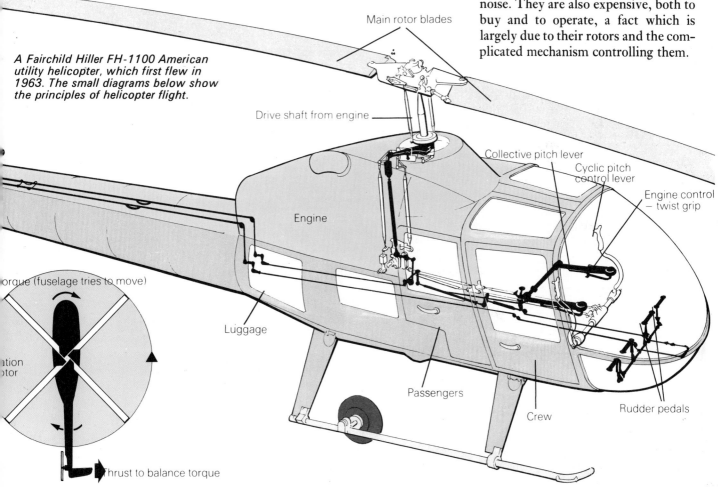

A Fairchild Hiller FH-1100 American utility helicopter, which first flew in 1963. The small diagrams below show the principles of helicopter flight.

Balancing the torque

HELP FROM THE AIR

Above: *A Fletcher FU-24, built by New Zealand Aerospace Industries Ltd, spreading phosphate fertilizer on barren hills in New Zealand.*

During the 1920s, farmers in the United States took to the air to wage war on insects which were eating their crops. Old biplanes flew low over the fields, dropping a stream of powder which fell on the crops and killed the insects. *Dusting* crops from an aircraft had two advantages over the more conventional tractor: it was faster and it did not damage the crops with its wheels. The dropping of fertilizer or *top dressing* was developed in New Zealand where barren hills were converted to grass pastures for grazing sheep by dropping the much needed phosphate fertilizer from the air. In the United States, farmers pioneered the dropping of seeds from the air and large areas were sown with wheat and rice: in fact almost all the North American rice crop is now seeded from the air. Liquid pesticides could also be sprayed onto the crops very easily by air.

The early agricultural aircraft were all second-hand machines converted for their new work. This involved fitting a large tank or *hopper* into the fuselage and mounting the spraying or dusting gear along the wings. The aircraft converted in this way ranged from small de Havilland Tiger Moths to Boeing B-17 Flying Fortress bombers. Helicopters were also used and they had two advantages over other aircraft: they could operate in confined spaces such as narrow valleys, and the 'downwash' from their rotor blades helped to blow the chemical right into the heart of the crop. During the 1960s, farming by air became a very successful business and aircraft were specially produced for such hard work. The resulting aircraft were not very elegant because they had to carry all the external spraying gear and be strong enough to stand up to frequent landings on rough ground in order to recharge the hopper. The cockpits were carefully designed to keep out the chemicals and to withstand a crash – accidents do happen, especially when the spraying 'runs' are made a few feet or so above the ground. Many of the successful agricultural aircraft have been built in the United States where they are widely used. The New Zealand inspired American Fletcher FU-24 was one of the first to sell in large numbers but there were a number of American rivals such as the Piper Pawnee, the Cessna Agwagon and the Grumman Ag-Cat biplane.

Surveying from the air

An observer in an aircraft obtains a 'bird's eye view' of the Earth which can be very useful, and, in addition, it can if necessary be recorded with a camera. Balloons and military aircraft have been used for observation purposes since their early days, and in the 1920s aerial map-making of the Earth's surface started. With a camera pointing downwards from the aircraft, photographs were taken at regular intervals along a straight line. By timing the camera carefully the photographs were taken so that they overlapped slightly. Having completed one run, the aircraft returned alongside the original line, and by flying to and fro in this way the required area was covered by a checkered pattern. Modern map-makers still operate in the same way but their equipment has improved tremendously.

Maps usually show what is visible on the surface of the Earth, but geologists are interested in the materials beneath the surface. Aircraft are equipped as flying laboratories, carrying delicate instruments which record information continuously during a survey flight. From these records the geologists can find out if there are any valuable minerals such as oil, uranium or iron ore beneath the Earth's crust.

Even historians have taken to the air to study archaeological remains. Patterns of buildings, long since demolished, can often be seen from the air. Fields of oats reveal remains of old buildings particularly clearly. Because the soil in the region of the building is often poorer than the surrounding area, the oats grow less well and so reveal the shape of the building. The Royal Air Force photographed Hadrian's Wall in the north of England as a training exercise and archaeologists discovered remains of several Roman camps which were previously unknown.

Below: *Archaeological remains seen from the air. Crops grow less well over buried foundations, revealing the shape of buildings.*

Farming from the air

Aircraft can be used in countless ways by an enterprising farmer – especially if he has a very large farm or ranch. Sometimes he flies around in his aircraft instead of driving a car or riding a horse. For example, a boundary fence can very easily be checked from the air to ensure that it has not been damaged. A modern cowboy might round-up his cattle by helicopter and fruit-growers find the downwash from a helicopter's rotor very useful for drying wet fruit. In the United States, the Fish and Wildlife Service has used aircraft to hunt wolves and to drop baby trout into lakes. But one of the most unusual jobs carried out by aircraft is to make it rain. In areas suffering from a long drought, clouds have been 'seeded' with chemicals and this has brought on the much-needed rain.

Right: *'Rain maker' aircraft in Australia.*

STOL AND VTOL AIRCRAFT

All aircraft have to fly slowly some of the time, especially during take-off and landing, but some aircraft are designed to fly slowly for long periods. This creates a problem because one of the most important factors in producing lift from a wing is speed. Reducing the speed reduces the lift, yet the wing still has to support the weight of the aircraft. To compensate, extra lift is needed, and the designers have to study the other factors affecting lift (*see page 14*). The two high-lift devices most widely used are: large extending flaps from the trailing edge of the wing; and slats along the leading edge which produce slots. Flaps increase the lift by changing the shape or camber of the wing, and they can increase its area. Slats and slots allow the aircraft to fly with a greater angle of incidence (nose up) without stalling.

Short take-off and landing aircraft

Special aircraft have been designed to operate from short landing strips in jungles or remote regions and these are usually referred to as *STOL* aircraft (Short Take-Off and Landing). One of the first such aircraft was the Handley Page Gugnunc of 1929 which had large

flaps and movable slats. The Handley Page Company had introduced a fixed slat and slot design in 1920, but the movable slats on the Gugnunc retracted onto the leading edge thereby closing the slot when the aircraft was flying quickly. The inboard slats were connected to the flaps and moved out from the wing to prevent stalling as the flaps were lowered to increase lift. The outboard slat sections moved automatically when the speed of the aircraft dropped and therefore delayed the stall.

During the Second World War, STOL aircraft were widely used for communications duties, and these included the German Fieseler Storch and the British Westland Lysander. Both could land almost anywhere with vital messages or important passengers. The Lysander was also widely used for landing spies, and supplies for resistance fighters, behind the enemy lines.

After the war several companies produced STOL aircraft for both civil and military use. For example, Scottish Aviation built their Pioneer and followed it with the larger twin-engined Twin-Pioneer. Both these aircraft made their mark flying from short airstrips hacked out of the jungle in the Far East. Canadian 'bush' pilots also have to operate from short, rough airstrips, consequently the Canadian company de Havilland Canada have concentrated on STOL designs. Their Otter was followed by the turbo-prop Twin Otter and more recently they have built the four-engined Dash 7 which can carry 50 passengers. The Dash 7 is described as a 'quiet STOL' airliner and it is designed to operate from small airports near city centers — so low engine noise is an important feature.

Below: *Britain's VTOL aircraft, the Harrier — a two-seat trainer version.*

Vertical take-off and landing aircraft

It took many years for aircraft designers to produce a really effective Vertical Take-Off and Landing, or *VTOL*, aircraft in order to dispense with large and costly airports. The helicopter fulfills some of the requirements: it can certainly take off vertically but the large rotor makes it difficult to produce a helicopter capable of high speeds; also helicopters are expensive and noisy. A number of strange-looking VTOL aircraft appeared, often using several small rotors (or large propellers) to lift the machine off the ground. Sometimes two or four engines were mounted on a tilting wing: for take-off the wing and engines pointed upwards so the propellers lifted the aircraft off the ground, then the wing was tilted into the usual horizontal position with the engines pointing forwards and the *convertiplane* flew as a conventional aircraft. None of these aircraft progressed beyond the experimental stage.

The invention of the jet engine with its powerful thrust made VTOL aircraft a distinct possibility. In 1953 the Rolls-Royce 'Flying Bedstead' lifted into the air, powered by two Nene engines with their jets directed downwards. The Short SC-1 followed, using four *lifting* engines and one more engine for forward propulsion, but this aircraft was only experimental – although its layout is favored by some people for the airliner of the future. Another variation is to change the direction of a jet between take-off and flight – either by tilting the engine or by deflecting the jet as it emerges from the engine. The most successful aircraft with this *vectored thrust* is the British Hawker-Siddeley Harrier fighter flown by the Royal Air Force and other airforces and also the US Marine Corps. The Harrier has a single engine but its jet emerges from four movable nozzles, which point downwards for take-off or to the rear in flight.

VTOL aircraft which are designed to carry passengers have one great advantage, in theory, for they should be able to operate from small areas of ground near a city center. In practice there are many problems: VTOL aircraft are very expensive to operate because hovering burns up large quantities of fuel, they are extremely noisy and the jet itself can cause havoc on the ground. The VTOL aircraft has so far failed to supersede the helicopter and future designs may produce a cross between the two to give an improved all-round performance. The Bell Helicopter Company of the United States has been experimenting with *tilt-rotor* machines since the late 1950s. By the

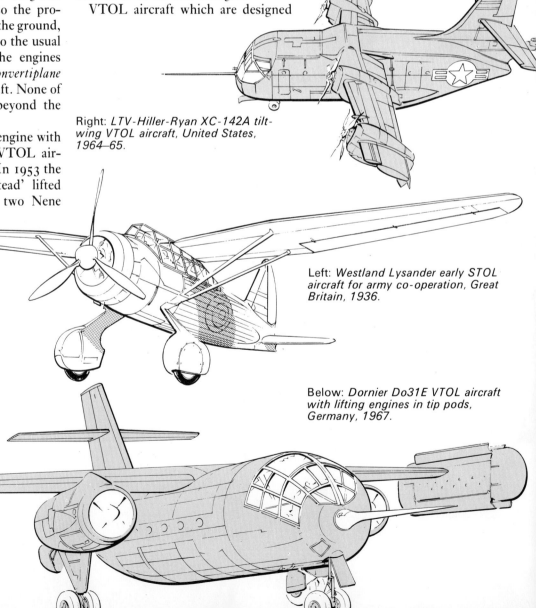

Right: *LTV-Hiller-Ryan XC-142A tilt-wing VTOL aircraft, United States, 1964–65.*

Left: *Westland Lysander early STOL aircraft for army co-operation, Great Britain, 1936.*

Below: *Dornier Do31E VTOL aircraft with lifting engines in tip pods, Germany, 1967.*

early 1980s their XV-15 was still experimental. In normal flight it looks like a conventional twin-engined aircraft but its turbo-prop engines are mounted at the wing tips. For take-off these engines and their rotors pivot to point upwards. Bell claims that the XV-15 can travel twice as fast as a helicopter, yet its noise level is about the same as an average helicopter.

There have been some weird and wonderful designs for VTOL aircraft produced over the years but it may be many more years before they carry passengers and freight.

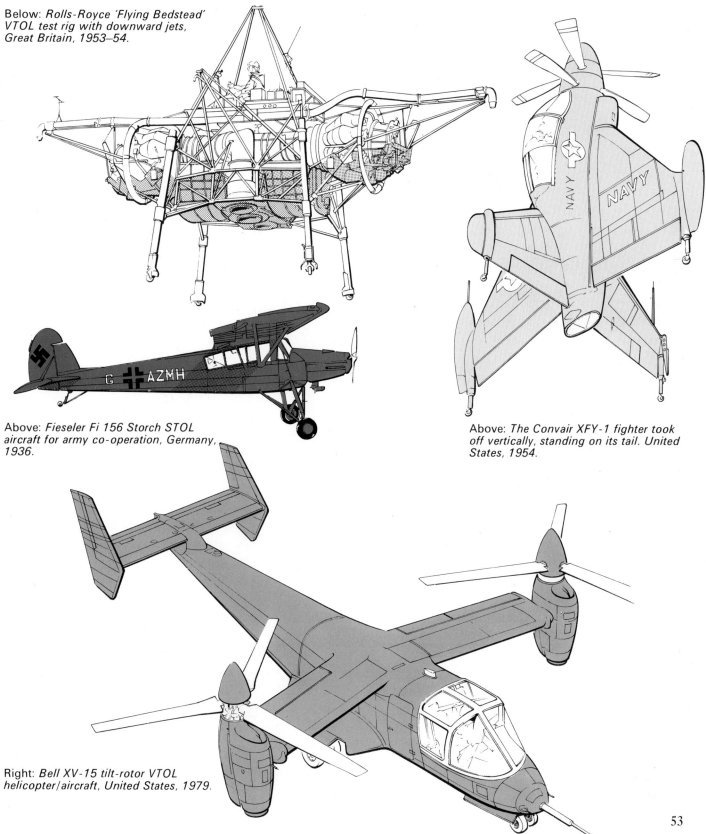

Below: *Rolls-Royce 'Flying Bedstead' VTOL test rig with downward jets, Great Britain, 1953–54.*

Above: *Fieseler Fi 156 Storch STOL aircraft for army co-operation, Germany, 1936.*

Above: *The Convair XFY-1 fighter took off vertically, standing on its tail. United States, 1954.*

Right: *Bell XV-15 tilt-rotor VTOL helicopter/aircraft, United States, 1979.*

A Junkers F13 airliner helping in the dramatic rescue of the survivors from the Soviet ship Chelyuskin *in 1934.*

AIR RESCUE

In an emergency, speed of action can make the difference between life and death, so, if long distances have to be covered, high speed aircraft are of great importance. Within the relatively short history of aviation there have been many examples of dramatic rescues and emergency relief. An early rescue took place in 1934 when the Russian steamer *Chelyuskin* sank in the Arctic. Despite many difficulties, aircraft landed on the ice and many people were flown to safety.

When natural disasters such as floods, earthquakes, hurricanes, or famines occur, aircraft are almost always on hand to rescue stranded people and fly-in food supplies. For example, farms and villages in mountainous areas can easily be isolated by a heavy snowfall and if food supplies run low, lives could be in danger. Aircraft can drop supplies of food for both people and animals.

When someone is seriously ill they need attention quickly which usually means getting a doctor to the patient or transferring the patient to hospital. In a village, town or city this is not normally a problem but in a very remote area, fast travel becomes vital. In some parts of Australia, Africa and South America there are settlements where the only means of reaching civilization used to be a long ride on horseback. Aviation has provided the answer to this problem and they now have flying doctor services.

Australia has one of the best schemes, called the Flying Doctor Service when it was founded in 1927 (becoming the Royal Flying Doctor Service in 1955). The problem it had to solve was twofold – poor surface transport and no telephone system. To improve communications, isolated farms were equipped with radio transmitter-receivers, but many of them did not have any electricity either so they had to make their own by driving a generator fitted with bicycle pedals. Once the farmers could radio for help, aircraft were hired from the local airlines to fly out a doctor or transport the patient to hospital. Aerial ambulances are also used in Britain to rescue patients in the 'highlands and islands' of Scotland. A normal airline service can easily be diverted to pick up a patient in an emergency or a special helicopter flight might be ordered if the patient is stranded on a mountain or at sea.

Above: *A United States Air Force pararescueman with his equipment prepares to jump.*

In the United States a new rescue service has been introduced in recent years using 'para-rescuemen'. These highly trained men drop by parachute to help anyone in trouble, but one of their most important tasks has been to recover space capsules and the astronauts after their journeys into Space. The para-rescuemen have to be trained in many skills including parachuting, first aid, mountaineering, arctic survival and skin diving.

Below: *A DHA Drover ambulance aircraft brings in a patient from the outback in New South Wales, Australia.*

Fire-fighting by air

Aircraft can be used as mobile viewpoints, which makes them ideal for spotting forest fires. Trees are a valuable crop, grown to produce timber and wood pulp for paper-making, and a forest fire can destroy them only too quickly. A fire which gets out of hand can also endanger human and animal life, so the effects of a forgotten camp fire or of lightning can be devastating.

In 1919 the United States Army Air Corps began to patrol some forests in California and from this small beginning, world-wide fire-spotting operations have grown. Pilots soon learned to spot fires, almost before they happened, by keeping an eye on campers and following in the path of any dangerous-looking thunderstorm. In the 1930s fire-fighting equipment was first dropped from the air to firemen on the ground. Then in the 1940s firemen were dropped by parachute in the region of a fire. This was rather dangerous but the 'smoke-jumpers' got to the fire very quickly and, with luck, they could bring it under control before it had time to spread and become really dangerous. Since the Second World War helicopters have been introduced to carry firemen and their equipment to forest fires – or any other fire in a remote area.

Water-bombing

A new method of fighting forest fires has been widely used in recent years and this is called water-bombing. Each aircraft is fitted with a large water tank and a release valve so the pilot can drop the water on a fire. One load is not usually sufficient, so the pilot has to fill up his tank and return to the fire. Seaplanes are frequently used because there are usually lakes to be found in forest country – but no airfields. With a seaplane the pilot can alight on the lake and fill the tanks by scooping up water as the aircraft taxies into position for a quick take-off.

The Canadian company Canadair designed an aircraft specifically for water-bombing. Their CL-215 is a twin-engined, amphibious flying-boat which has very large tanks, mounted in its fuselage, capable of carrying 1420 gallons of water – 10 times as much as earlier water-bombers. What is even more remarkable is the fact that these tanks can be filled, as the CL-215 skims across the water, in under 20 seconds.

PEOPLE AT WORK

Around the world there are hundreds of thousands of people whose work depends upon aviation. Many work directly for an aircraft manufacturing company, for an airline or at an airport, and some of these jobs have been mentioned in earlier sections. But there are many other people, in a wide variety of industries, who are less obviously dependent upon aviation for a livelihood.

Before an aircraft can be built, the manufacturer needs the basic materials. Wood has been replaced by metals, and in particular aluminium and its alloys (aluminium with a small amount of copper, zinc or magnesium added). Even after the discovery of these alloys, the metal companies had to develop methods to produce them at reasonable prices. Steelworkers are also involved in the aviation industry, for steel is used where its high strength is required on such items as engines and undercarriages. Tires are also needed and they are supplied by the rubber manufacturers. Instruments are supplied by specialist firms, while furnishing manufacturers, glass makers and many others all contribute to the success of an aircraft.

The aero-engine companies have built up a whole industry to design, develop and manufacture engines for a wide range of aircraft: they have even adapted their engines for other uses. For example, a modern frigate in the British navy is powered by two Rolls-Royce Olympus together with two Rolls-Royce Tyne gas turbine engines. These are marine versions of the aero-engines used in Concorde and the Vickers Vanguard. Another use for modified aero-engines is to pump North Sea gas along the long pipe-lines to the areas where it is needed.

Engines need fuel and oil, so most of the large international oil companies have specialist sections which produce and distribute aviation fuel to the four corners of the world. At major airports each company has its own staff to refuel aircraft – either by tanker vehicles or a piped-hydrant system. In contrast to these operations, fuel has also to be available in remote areas such as the Australian 'outback' or the Canadian 'bush'. The owner of a local garage or a general store may sell aviation fuel as part of his business.

Research and development in the oil industry are vital; fuel technologists are constantly striving to improve their products, and maintain the quality of existing supplies. Should any impurities such as water or dirt find their way into aviation fuel, the result could be an engine failure and a major catastrophe. The oil companies are not only suppliers to the aviation industry – they are also good customers. Oil wells and pipe-lines are usually situated in remote areas where surface travel is difficult, consequently air transport is used. For example, most of the oil rigs in the North Sea have a regular helicopter service linking them to the mainland.

Below: *Tires are a vital component which have to be checked regularly for damage.*

Right: *Height-measuring altimeters are tested and inspected at each stage of manufacture.*

Servicing and research

Keeping a single aircraft in the air can involve hundreds of workers on the ground, particularly if it is a large airliner or a powerful military aircraft. Routine servicing is usually carried out on the spot, but major overhauls or modifications are often handled by specialist companies. Modifications may involve converting an old airliner into a freighter, or converting a military aircraft to take new equipment.

When an aircraft is serviced at an airport it is surrounded by a host of specialized vehicles and equipment. For companies who manufacture tanker vehicles, tractors, trailers, fork-lift trucks, runway sweepers, special buses and other mechanical equipment, the aviation industry is a large and lucrative customer.

Few industries move forward faster than the aviation industry and many people are engaged in research. This ranges from developing new materials to the theory of flight, particularly supersonic flight. Work is carried out by government-sponsored establishments, universities and manufacturers. Sometimes military aircraft are used to assist with research projects. For example, in the USA a considerable amount of research into supersonic flight was carried out by the United States Air Force in conjunction with the National Aeronautics and Space Administration. Britain's Royal Air Force carried out a series of flights in the region of the north magnetic pole which assisted in the solving of the navigational problems in that area.

In addition to research projects, government establishments employ many scientific, technical and legal experts to maintain the safety standards of aircraft based in their country. The electronics industry also works very closely with the aircraft designers to supply radio, radar, navigational aids and other equipment.

Below: *A Boeing 707 being serviced.*

LIGHT AIRCRAFT

For business people 'time is money' and this includes traveling time which, as far as the company is concerned, is wasted time. After the Second World War, the use of business aircraft grew, steadily at first, but then quite rapidly.

In the United States, where long distances are a problem, business aircraft are widely used. Many senior executives have a jet aircraft converted for business use, complete with desk, telephone, typewriter, conference table and even a bar. Some extremely wealthy executives and oil millionaires own jet airliners converted for their personal use with luxurious fittings including a bed and even a bathroom. Most executive aircraft, however, are more modest, carrying from four to twelve passengers.

In Britain, de Havilland's Dove and Heron of the late 1940s were widely used as executive aircraft and they were succeeded by the small twin-jet Hawker Siddeley HS-125 (de Havilland being incorporated in Hawker Siddeley by then). The HS-125 can cruise at 500 mph (800 kph) and it has even sold in America – the home of the business aircraft! The choice of aircraft in America is very wide and competition for sales is fierce. For example, one company, Cessna, offers over 30 different models ranging from the two-seater Model 152 to the luxurious Citation powered by two jet engines. Their great rivals, Beechcraft and Piper, also offer a similar selection, and a number of other American companies offer executive jets. Although the sales of executive aircraft are dominated by the American manufacturers, there are manufacturers in many parts of the world. From Canada comes the Canadair Challenger, while in France Dassault-Breguet produce the Falcon, and Aérospatiale the Corvette. Mitsubishi of Japan build the Diamond, and Embraer of Brazil manufacture the Xingu.

Below: *Cessna Model 172 four-seat light aircraft – one of the most popular aircraft of all time; United States, 1955.*

Below: *De Havilland DH104 Dove Series 1 small airliner and executive aircraft, Great Britain, 1945.*

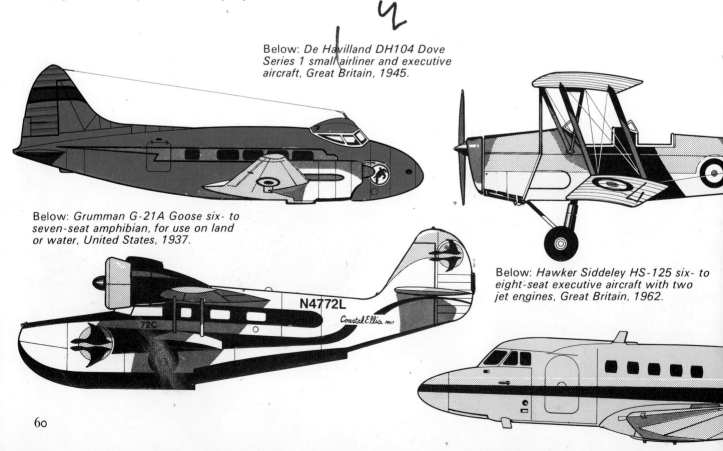

Below: *Grumman G-21A Goose six- to seven-seat amphibian, for use on land or water, United States, 1937.*

Below: *Hawker Siddeley HS-125 six- to eight-seat executive aircraft with two jet engines, Great Britain, 1962.*

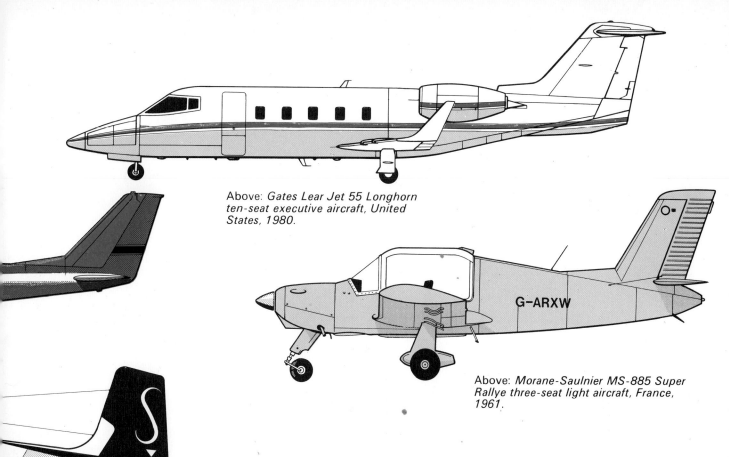

Above: *Gates Lear Jet 55 Longhorn ten-seat executive aircraft, United States, 1980.*

Above: *Morane-Saulnier MS-885 Super Rallye three-seat light aircraft, France, 1961.*

Above: *Cessna Model 310B five-seat twin-engined private or light business aircraft, United States, 1953.*

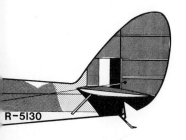

Above: *De Havilland DH82 Tiger Moth biplane trainer in wartime markings, Great Britain, 1931.*

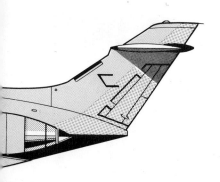

Private aircraft

In the years before the First World War, flying as a hobby was only for people with plenty of money, and the average member of the public just went along to watch. Things changed after this war when there was a determined effort to get more people into the air: joy rides, flying clubs and home-built aircraft were all available at modest prices. Flying clubs were founded near most large towns and many light aircraft were built for amateur pilots: these could be bought for little more than the price of a car. One of the most popular light aircraft of this period was the de Havilland Moth biplane which was followed by the Gipsy Moth and the Tiger Moth.

The Second World War brought private flying in Europe to a halt. When peace returned in 1945 some of the clubs re-emerged but the cost of flying in Britain has never decreased to the level of the 1930s. In the United States, the story is very different, for private aircraft are extremely popular and flying is relatively inexpensive. The term *General Aviation* (G.A.) is used to cover private, business, air-taxi and all other branches of aviation other than military and airline operations. There are over 200 000 G.A. aircraft in the United States and the figure continues to rise. The modern equivalent of the de Havilland Moth is the American Cessna Model 150 and its successor, the Model 152. The two-seater Model 150 first flew in 1957 and by the time it was phased out in 1977, a total of almost 24 000 had been built. Even this success was surpassed by the larger, four-seater Model 172 Skyhawk for which the production figure passed 30 000. All three of these Cessnas are single-engined high-wing monoplanes which can cruise at about 125 mph (200 kph). They are widely used by private owners and flying clubs in many parts of the world.

Some private owners use their aircraft instead of a car – especially in the United States – but others fly as a hobby. For those who enjoy competitions there are air rallies, races and aerobatic competitions. The latter two sports can be very competitive and specially designed aircraft are produced for these events.

THE FUTURE

In Jules Verne's famous story *Around the World in Eighty Days* written in 1873, the only air travel mentioned was a short trip in a balloon; the rest of this eventful journey made use of practically every other form of transport from elephants to steamers. Today it is possible to sit in the comfort of an airliner and fly round the world in a few days.

Jules Verne would not have been surprised by this – indeed, many of the predictions made by writers of fiction, such as Jules Verne and H. G. Wells, have proved to be more accurate than the forecasts of acknowledged experts.

Over the years, experts have placed limits on almost every aspect of aircraft design and performance – only to be proved wrong a few years later. The maximum possible speed for an aircraft was once thought to be about 150 mph (240 kph), and in the 1920s a United States Navy committee decided that the largest possible aircraft would weigh about 44 000 pounds – a Jumbo weighs 770 000 pounds. So the designers of aircraft have often proved the predictions wrong, but sometimes the designers have been wrong too. Some of their major mistakes have been with large aircraft which were ahead of their time. Back in 1921 the Italian designer Count Gianni Caproni built a flying boat which had nine wings (three sets of three) and eight engines. It was in-

Left: *One of the most unusual aircraft ever built – Count Caproni's flying boat with nine wings, 1921.*

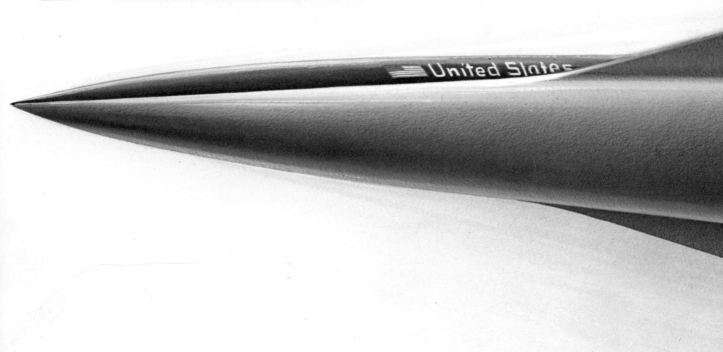

tended to carry 100 passengers but after two short hops the unwieldy machine was abandoned. History repeated itself several times over and one of the most spectacular repeats was Howard Hughes' giant Hercules (*see pages 44–5*).

Will the airliners of the future continue to increase in size or is the Boeing 747 Jumbo, with its 490 seats, the end of the line? Larger aircraft result in cheaper air fares so there is a great temptation to increase the size of airliners. On the other hand, the existing runways of the world's airports cannot cope with ever-increasing aircraft weights and these runways may prove to be the brake on progress. A group of some 500 passengers can present a problem when they have to be checked through passport control or moved to a distant loading gate. Consequently, larger groups would need improved passenger-handling arrangements – perhaps conventional passports will be replaced by plastic cards which could be checked by a computer.

Some routes just do not have enough passengers for the largest airliners so there will continue to be a use for the smaller airliner. For example, 500 passengers per day may wish to travel on one particular route. It would be economical for an airline to operate one flight per day using a 500-seat airliner, but this might not suit the passengers and five flights during the day, each carrying 100 passengers, might be more acceptable. There will certainly be a considerable effort to make all airliners, large or small, quieter, safer and less expensive to operate. Because fuel is so expensive, engines must be designed to use less fuel. On many of the shorter routes, helicopters, STOL and VTOL aircraft will probably dominate the scene for some years to come, but the future of the flying boat is doubtful.

Below: *Artist's impression of a future airliner powered by hydrogen and capable of 3900 mph.*

Faster or slower?

Will airliners of the future fly faster or slower than those of today? People are always looking for ways of saving time – the faster we travel, the more time we save. But there are problems. Conventional subsonic airliners cannot fly any faster without running into the high drag of the 'sound barrier'. Supersonic airliners, such as Concorde, are expensive to operate – although they may become cheaper in the future. But an airliner designed to fly much faster than Concorde would be very expensive to produce because it would have to be built from stainless steel and titanium to withstand the high temperatures generated at Mach 3 and above. It has even been suggested that passenger-carrying space rockets might be built to transport passengers between Britain and Australia in only an hour or so.

Of course, speeds may change in the other direction, with slow-flying air travel gaining popularity. The airship might return, perhaps with nuclear-powered engines. These engines could run for years on just a drop of fuel, but the protective shield around the reactor would need to be so heavy that, at present, the prospects do not look too promising. A new development in nuclear science could change this state – or an exciting new source of power might be invented which would reduce the cost of air travel to a fraction of its present figure.

For most passengers, hopes for the future are largely directed towards reliable, safe and cheap air travel.

The airliner of today – but what does the future hold in store?

INDEX

aerial: archaeology, 49, *49;* farming, 49; rainmaking (Austr.), *49;* rescues, *46,* 47, 54–56, *54–55;* surveys, 49; waterbombing, 57, *57*
Aer Lingus, 20, *38*
aerobatics, 61
aerodynamics, 14; loading, 28
aero-engine companies, 58
Aeroflot (USSR), 38, *38*
Aerospace Industry (NZ), *48*
Aero Spacelines 'Guppy', 40, *40*
Aerospatiale Corvette (Fr.), 60
Ag-Cat, Grumman aircraft, 48
agricultural aircraft, 48–49, *48, 49*
Agwagon, Cessna aircraft, 48
aileron, aircraft, 15, *22*
Airbus, *15–40;* Airbus Industrie A300, 21; A300B, *22–23;* testrig, *29*
Air Canada, *38,* 39
aircraft: flight principles, 14–15, *14–15,* 50
aircraft, types: amphibian, 42, 44, 57; floatplanes, 31, 42, 44, *45;* flying-boats, 11, *11,* 13, 17, *17,* 42–45, *42–45,* 63; pickaback, 45, *45;* seaplane, 42, 57; STOL/VTOL, 50–53, *50–53*
aircraft, uses: agricultural, 48–49, *48–49;* firefighting, 56, *57;* freight, 40–41; military, 49; private, 60–61; *60–61;* rescue service, *46,* 47, 54–56, *54–55;* surveys, 49, *49*
Air France, 16, 38, *38*
Air Fret (Fr.), 38
Air India, *38,* 39
Air Jamaica, *38*
airliner, 8–11, 17, *17,* 18–19, 28–29, *28,* 30–31, 62–63; jet, 20–27; stack, airport, *31*
 see also Airbus
airlines, 16–17, 38–39, *38–39*
airmail, 9, 16, 40
airport, *30–31,* 33, 34–37, *34–35,* 36, 37, 42, 59, *59*
airship, 4–7, *4, 5–7, 7*
Air Traffic Control, 30, *30,* 31, 33; airways, 32–33, *32–33*
air travel, 16–27
Akron, airship, 5
Alcock, Captain John, 5
Alitalia (It.), *38*
altimeter, 58, *58*
aluminum, 28, 58
American Airlines, 38; American Airways, *38*
amphibian aircraft, 42, 44, 57; Goose, *60–61*

Amundsen, Roald, 11, *11*
Andrée, Salomon, North Pole, 3, *3*
animal, air cargo, 41
Antarctic, 11
Apollo spacecraft: 40
Arctic, aerial rescue, *54–55*
Arctic, *3,* 11
Argosy, airliner, 16, *16,* 17
Aristotle, 14
Armstrong Whitworth, Argosy, 16
Army Air Corps, 56
Atlantic: air crossing, 3, 5, 10, *10,* 17, 33, 42
Australia: aerial rainmaking, *49;* air routes, 10; aviation fuel, 58; Royal Flying Doctor Service, 56, *56*
autogiro, 46, *46*
automation: aircraft manufacture, 29
autopilot, aircraft, 31
Avro Vulcan bomber, *27*

BAC One-Eleven, jet aircraft, 20
balloon: 2–3, *2,* 14; Andrée's, 3, *3;* observation, 49; Verne, 62
Beechcraft, 60
Bell Helicopter Co., 52
Bell X-1 research aircraft, 25, *25*
Bell XV-15 VTOL aircraft, 52–53, *53*
Bennett, Floyd, 11
Benz, Karl, 5, 12
Berlin Air Lift, 41, *41*
biplanes, 9
Blanchard, Jean-Pierre, 2
Bleriot, Louis, 9, 42
blimps, 6
Boeing: B-17 Flying Fortress, 48; 314 flying-boats, 42; 747 Jumbo jets, 13, 19, 21, 22–23, 25, *25,* 37, 39, 63; Model 40, 40; 377 Stratocruiser, 19, 40; Boeing 707, 20–21, 22, 25, *25, 37,* 38, 39, *59;* B727, 21
bogies: Jumbo jet, 22
bomber, 5, 9, *10,* 13, 16, 20, 48
brake: hydraulic, Jumbo jet, 23
Brand, Quintin, 10
Brazil, Xingu, 60
Breguet, Louis, 46
Bristol Aero-engine Co.: 13; 170 Freighter, 40; Hercules, 13, 20; Olympus, 13
Britain: Airbus A300B, *22–23;* Civil Aviation Authority, 35; supersonic aircraft, 26–27, *26–27;* VTOL aircraft, 50–51
British Airways, *32–33,* 38, *38*
British Caledonian, *38*
British European Airways, 20

British Overseas Airway Corp, *18*
Brown, Lieut. Arthur Whitten, 5
bush pilot (Can.), 41, 44, 51, 58

Calcutta, Short flying boat, 42
Canada: bush pilots, 41, 44, 51, 58
Canadair, *CL-215,* 57, *57*
Cape Kennedy, 40
Caproni, Count Gianni, 62–63; flying boat, *62*
Caravelle (Fr.), 20, *20*
cargoes: air freight, 40–41, *40*
Catalina, Consolidated PB7-5, 44
Cathay Pacific Airways (Hong Kong), *38*
Cayley, Sir George, 8, 46
Cessna: 60; Agwagon, 48; Citation, 60; Models 150/152, 61; Models 172 Skyhawk & 320B, *60–61*
Challenger (Can.), 60
Chanute, Octave, 8, 28
Charles, Professor Jacques, 2
chart: Airways Navigation, *32–33*
Chelyuskin aerial rescue, *54–55*
Cierva, Juan de la, 46, *46*
Citation, Cessna, 60
CL-215 (Can.) amphibian, 57, *57*
coal: gas, 3
Cobham, Sir Alan, 11, *11*
collective pitch level, helicopters', *47*
Comet (GB), 20, *21,* 25, *25,* 26, 29
computer: aircraft, 30; air traffic control, 33
Concorde (GB/Fr.), 13, 21, 25, *25,* 26, *26–27,* 28, 38, 58
Constellation, Lockheed aircraft, *18, 19*
container, air cargo, 41
control column, aircraft, 15
Convair 880/990, 21; XFY-1 fighter, *52–53*
Cornu, Paul, 46, *46*
Corvette (Fr.), 60
Coxwell, 2
crew, airliner, 30–31
Croydon Airport, 36, *36*
Curtiss NC4 floatplane, 31, 42, *44*
cyclic pitch control, helicopter, *47*
cylinder: air, 12–13, *12*
Czechoslovak Airlines, *38*

Daimler, Gottlieb, 5, 12
Dakota DC-3, 18, *18,* 19, 28, 38, 40
d'Arlandes, Marquis, 2, 3
Dart turbo-prop engine, 20
Dash 7 STOL (Can.), 51
Dassault-Breguet (Fr.) Falcon, 60

Decca, navigator, 32
de Havilland Aircraft Co., 20, *21*, 48;
  Comet, 25, *25*, 26, 29; DH82
  Tiger Moth, *60–61*; DH104 Dove,
  60, *60–61*; Gipsy Moth, 61;
  Heron, 60; Tiger Moth, 48, 61;
  Trident, 20
de Havilland Canada, STOL, 51
delta wings, Concorde, 26
Deutsche Lufthansa (Ger.), 16, 42
DH82 Tiger Moth (GB), *60–61*
DH104 Dove, *60–61*
Diamond (Jap.), 60
diesel engine: Jumbo, 13; marine, 6, 13
Doctor Service, Royal Flying, 56, *56*
Doppler, radar, 32
Dornier DO31E VTOL (Ger.), *52–53*;
  DOX, 44; Wal, 42
Double Eagle 11 balloon, 3, *3*
Douglas: DC-3 Dakota, 18, *18*, 19, 28,
  38; D-4, 18; DC-7C, 19; DC-8,
  21; DC-9, 21; DC-10 jet, 13, 21,
  28; World Cruisers, 10
Dover, 2
Dumont, Alberto Santos, 4, *4*
duralumin, 28

Eastern Airlines, 38
Eckener, Dr. Hugo, 5
Electra, Lockheed aircraft, 20
elevator, aircraft, 15, *22*
Embraes (Braz.) Xingu, 60
engine, power: diesel, 6, 13; electric,
  4–5, 12; gasoline, 4, 5, 8, 12, 19;
  gas-turbine, 58; nuclear, 6;
  steam, 4, 8, 12
engine types: internal combustion,
  4–5, 8, 12–13; jet, 13, *13*,
  19–27, 52
executive aircraft, 60, *60–61*

Fabre, Henry, 42
Fairchild Hiller FH-1100 helicopter,
  46–47
Falcon (Fr.), 60
fans, turbo, 13
fatigue failure, aircraft, 29
Fieseler Storch STOL (Ger.), 51,
  52–53
fire-fighting, aerial, 56–57, *57*
First World War, 5, 13, 28, 61
Fletcher FU-24 (NZ), 48, *48*
Flettner & Kolibri (Ger.) helicopters,
  46
flight principles, *14–15*, 50
floatplane, Curtiss, 30, 32, 44, *45*
Flyer, 8; *Flyer No. 3*, 9, 28, 42
Flying Bedstead VTOL (GB), 52,
  52–53
flying boat, 11, *11*, 13, 17, 42–45,
  42–45, 63
Flying Doctor Service, Royal, (Aus.),
  56, *56*

Flying Fortress bomber, 48
Flying Junkers, 13
Fokker FV11, 10, 11
Fokker F27 Friendship (Neth.), 20
France: Airbus Industrie A-300, 21,
  22–23; Concorde, 13, 21, *25*,
  26–27, 28, 38, 58
freight: air, 40–41, *40*
freighter, aircraft, 40–41, *40*
Furnas, C.W., 9

Galaxy, Lockheed aircraft, 40, 41
gas: types, 2–3; helium, 3
gasoline engine, 4–5, 8, 12, 19
Gatty, Harold, 10
Gee, navigation, air, 32
General Aviation, 61
General Motors, 13
geological survey (aerial), 49
Germany: Airbus A-300, 21, *22–23*, 29
Giffard, Henri, 4, *4*
Gipsy Moth, de Havilland, 10
Glaisher, 2
glider, 8, *8*, 12, 28; hang-glider, 8
Gnome rotary engine, *12*
Goodyear Tire Co. blimps, 6, *7*
Goose, Grumman amphibian, *60–61*
Graf Zeppelin, 5
Grumman: Ag-Cat aircraft, 48; G-21A
  Goose amphibian, *60–61*
Gugnunc (Handley-Page) aircraft,
  50–51
Gulf Air, *39*
Guppy freighter, 40, *40*
gyroscope, 31

Handley-Page aircraft (GB), 9, *9*, 17,
  17; Gugnunc, 50–51; *Helena*,
  HP42 airliner, *17*; slats, 15, *15*,
  50–51
Handley Page, Frederick, 15
Harrier VTOL (GB), *50–51*, 52
Hawker Siddeley (GB), 21; Harrier,
  50–51, 52; HS-125 jet, 60,
  60–61; Trident, 21
Heathrow Airport, London, 33, *34–35*
Heinkel (Ger.) HE178 engine, 13
Helena HP42, Handley Page
  airliner, *17*
helicopters, 46–49, *46–47*, 52, 56,
  58, 63
helium gas, 3
Henson, William S., 8
Hercules Bristol aero-engine, 13
Hercules flying-boat, *42–43*,
  44, 45, 63
Hercules, freight aircraft, *40*, 41
Heron, de Havilland (GB), 60
Hindenburg (Ger.), 6, *6–7*
Hinkler, Bert, 10
HS-125 jet aircraft, 60, *60–61*
Hughes, Howard, *42–43*, 44, 63
hydraulic: system, 23
hydrogen, 2, 3, 6

IATA (International Air Transport), 39
ICAO (International Civil Aviation), 39
Imperial Airways (GB), 16, 17, 42
industry: aviation, 58–59
inertial: navigation system, 32
insignia, airline, *38–39*
instrument landing system, 31
internal combustion engine, 12, 13
Iran Air, *39*

Japan: airlines, *39*
Jeffries, Dr. John, 2
jet: aircraft, 20–27, 60, *60*; engine,
  13, *13*, 19, 20–27, *22*
Johnson, Amy, 10, *10*
Jumbo jet *see* Boeing
Junkers: F13 *Chelyuskin* rescue,
  54–55; Jumo diesel aero-engine,
  13

Kingsford-Smith, Charles, 10
Kipfer, Paul, 3
KLM, 16, 39
Kolibri *see* Flettner
Krebs, Arthur, 4

La France, airship, 4–5
Langley, Samuel Pierpont, 8
Latecoere 521 (Fr.), 44
Lear Jet Model 23, *60–61*
light aircraft *see* private aircraft
Lilienthal, Otto, 8, *8*, 28
Lindbergh, Charles, 10
liner: passenger, 30
Lockheed: Constellation, 18, *18–19*,
  19; Electra, 20; Galaxy, *40*, 41;
  Hercules, *40*, 41; Star lifter, 41;
  Starliner, 19; TriStar jet, 13, 21
London: Air Traffic Control Centre,
  33; air travel, 17; Heathrow,
  33–34, *34–35*
Loran, aircraft navigation, 32
Los Angeles: International Airport, 37
LTV-Hiller-Ryan VTOL aircraft, *52–53*
Lufthansa (Ger.), *22–23*, 23, 38, *39*
Lysanders, Westland (GB), 51, *52–53*

Mach, Ernst, 24
Mach 1–2, 24, 25
Macon, airship, 5
map-making, aerial, 49
marshaller, airport, *30*, 33
Martin 130, flying boat, 42
Mayo, Major R.H., 45, *45*
Mediator, Air Traffic Control
  computer, 33
Merlin engine, *12*
Model 40 Boeing, 40
Montgolfier, E. and J., 2, *2*, 3
Morane-Saulnier (Fr.), *60–61*

navigation: aircraft, 32–33; chart,
  32–33
Nene jet engine, 20, 52
Nene-Viking, Vickers aircraft, 20

New York: airport, 33, 34
New Zealand: Aerospace Industry, *48;*
    Fletcher FU24, 48, *48*
Nigeria Airways, *39*
Nobile, General Umberto, 11
non-rigid airship, 5, 6
Norge, airship, *11*
North Pole, balloon flight, 3, *3*
North Sea: gas, 58; oil, 58

oil: companies, 58; North Sea, 58
Olympic Airways, *39*
Otter STOL (Can.), 51

Pacific crossing, 10; flying-boat, 17, 42
Pan American World Airways, 38,
    39, 42
pararescue service, 56, *56*
Pawnee agricultural aircraft, 48
photography: aerial, 2
Piccard, Professor Auguste, 3
pickaback aircraft, 45, *45*
Pilcher, Percy, 8, 28
Pioneer (Scottish Aviation), 51
Piper: 188; Pawnee agricultural
    aircraft, 48
Post, Wiley, 10
Pratt & Whitney, engine, 20
pressure, aircraft cabin, 22, *22–23*
private aircraft, *60–61, 60–61*
propane gas, 3
prop-jet engine, 20
pure-jet engines, 20

QANTAS (Austr.), 39, *39*

R33, R34 airships (GB), 5, *4–5,* 10;
    R100-1, airships, 5
radar: air, 31, 32, 33
radial engine, 12–13
radio: air, 32, 33; marine, 32–33,
    32–33; Navigation Chart, air,
    32–33
rain-making, aerial, *49*
Renard, Charles, 4
rescue service, aerial, *46,* 47, 54–56,
    54–55
research: aviation, 59
Richet, Professor, 46
rigid airships, 5
Robert, brothers, 2
Rolls Royce: 13; engine, *12,* 20, 26, 27,
    52, 58; Flying Bedstead VTOL,
    52, *52–53*
rotary engine, 12–13
rotor, helicopter, 46–47, *47*
routes: airways, 32–33
Royal Dutch Airlines (KLM), 16, 39
Royal Navy (GB) aero-engines, 58
Rozier, Pilâtre de, 2, 3

rudders: aircraft, 15, *22*
Ryneveld, Pierre van, 10

Sabena Airline (Belg.), 16, 39
Saudi (Saudi Arabia), *39*
Saunders-Roe, Princess, 45
SC-1, Short VTOL, 52
Scandinavian Airlines, 39
Schneider Trophy, 42, *45*
Scottish Aviation, Pioneer, 51
Sea King helicopter, *46*
seaplane, 42, 57
Second World War, 6, 9, 13, 18, 20, 24,
    28, 32, 40, 42, 44, 51, 61
Short: Empire flying-boat, 17, 42;
    SC-1, 52; Sunderland, 44
Sikorksy, Igor, 46, *46,* 47
Sikorsky, 42; VS300 helicopters, 46,
    46, 47
Silver City Airways, 40
Singapore Airlines, 39, *39*
Skyhawk, Cessna aircraft, *60–61*
slats, aircraft, 15, *15,* 50–51
Smith, Keith and Ross, 10
Snecma Olympus engine, 26, 27
sonic boom, supersonic aircraft, 24
South Africa: Imperial Airways, 17;
    BOAC jet, 21
Southern Cross, 10
Sperry, Lawrence, 31
Spitfire (GB), 13
Spruce Goose, flying boat, *44*
stack, airliner, *31*
Starlifter, Lockheed aircraft, 41
Starliner, Lockheed aircraft, 19
steam: engine, 4, 8, 12
steel: airplane, 28, 58
STOL (Short Take-off and Landing
    aircraft), 50–53, *52–53,* 63
Stratocruiser, Boeing, 19, 40
Stringfellow, John, 8
structure, aircraft, 28–29, *28*
Sunderland, Short flying-boat, 44
Super Guppy freight plane, 40, *40*
Supermarine, S.6B, 42, *45*
supersonic airliner, 21, 25, 28
Super VC10, Vickers aircraft, *21*

Tailplane, aircraft, 15, *22*
Tiger Moth, de Havilland, 48
tire: aviation, 58, *58*
titanium, 28
trans-polar flight, *11;* balloon, *3*
Trans-World Airlines, 38, *39*
Trident, airliner, 20, 21
TriStar, Lockheed jet, 13, 21
tunnel: wind, 24, *24*
Tupolev Tu-144 (USSR), 21
turbine: gas, 58
    see also jet engine

turbo engine, 13, 20; turbo-fan, 13,
    22; turbo-prop, 40
Twin-Otter STOL (Can.), 51
Twin-Pioneer (Scottish Aviation), 51

United Airlines, 38
United States: aerial farming, 48, *49;*
    airlines, 16, 18, 38, 39; Army Air
    Corps, 56; freighter, *40*

Vanguard, Vickers aircraft, 20, 40, 58
van, transporter, 56, *56*
Varig (Brazil), 39, *39*
VC10, Vickers aircraft, 20
Verne *(From the Earth to the Moon),*
    62
Vickers Armstrong Aircraft Co., 20
Vickers: Nene-Viking, 20; Super
    VC10, *21;* Vanguard, 20, 40, 58;
    VC10, 20; Viking, 20; Vimy
    bomber, 5, *10*
Viscount, *20;* Wellington bombers, 13,
    20
Viking, Vickers aircraft, 20
Vimy bomber, 5, *10*
Vinci, Leonardo da, 8, 46
Viscount, Vickers, aircraft, *20*
VOR/DME, navigation, 32, 33
VS 300 Sikorsky helicopter, 46, *46,* 47
VTOL, Vertical Take-off & Landing
    aircraft, *50–53,* 63

War: First World, 5, 13, 28, 61; Second
    World, 6, 9, 13, 18, 20, 24, 28, 32,
    40, 42, 44, 51, 61
Washington DC: airport, 37
Wasp engine, 13
water-bombing, aerial, 57, *57*
Wellington bomber, 13, 20
Wells, H.G. *(First Men in the Moon,
    The),* 62
Westland Lysander STOL (GB), 51,
    52–53
wheels: Jumbo jet, 22
Whitney *see* Pratt
Whittle, Sir Frank, 13
wings, aircraft, 14, *14–15, 22,* 25,
    29, 50
Wolfert, Karl, 5
World Cruiser, Douglas, 10
Wright engine, *12,* 13
Wright, Orville & Wilbur, 8, *9,* 12, *12,*
    28, 42

XFY-1 Fighter, *52–53*
Xingu (Brazil), 60

Yeager, Captain Charles, 25

Zeppelin, Count Ferdinand von, 5
Zeppelin (Ger.), 5, 6